D0848588

Energy and Environment:
A Primer for Scientists and Engineers

Energy and Environment:
A Primer for Scientists and Engineers

Edward H. Thorndike
University of Rochester

ADDISON-WESLEY PUBLISHING COMPANY
Reading, Massachusetts · Menlo Park, California
London · Amsterdam · Don Mills, Ontario · Sydney

ISBN 0-201-07497-4
.BCDEFGHIJ-AL-79876

To Liz,
". . . You come, too."

Preface

The "environmental crisis" was officially recognized on Earth Day, April 22, 1970; the "energy crisis" became official with the Arab oil embargo of 1973. Following these events, there has been strong urging that the general public should become better informed and educated about environmental problems and about matters relating to energy. A substantial number of books on environment and a smaller number of books on energy have been forthcoming, all of them aimed at the lay reader. This book is not one of them.

I have no quarrel with the need for a public enlightened on energy and/or environment. My cause, however, is the need for *professional scientists and engineers* to become broadly knowledgeable in these areas. Increasingly, scientists and engineers are being called upon to give advice here; few of them, however, have adequate knowledge that extends beyond their own areas of specialization. It is the very nature of environmental and energy problems that there are extensive interconnections, and a broad knowledge of the "big picture" is at least as important as a detailed knowledge of any one aspect of it.

This book, then, is written for practicing scientists and engineers and (more particularly) for students preparing themselves for these professions. It is intended to introduce them to the field of energy and environment, and it assumes they will supplement this overview with a detailed study of those areas that they find of particular interest.

It is hoped that most of this book will be comprehensible to a college sophomore and at the same time not appear overly pedestrian to a graduate student or practicing scientist. To this end, some background material has been relegated to appendixes. The problems that accompany each chapter have been given sufficient range in difficulty to provide something for everyone. Minimum prerequisites for a useful reading of the book include one-year college courses in both physics and mathematics and a one-year college course in chemistry, biology, or geology. More important than the knowledge of specific subject matter is the ability to deal with new material from a scientific point of view.

The principal use of the book is as the main text for a course that (except at a few institutions) does not yet exist: a junior-level, one-semester course on energy and environment for science and engineering majors. Lacking such a course at your institution, you can use the book as supplementary reading in courses on thermodynamics, nuclear physics, nuclear engineering, energy conversion processes, etc.

The topics united in this book under the heading "energy and environment" are normally assigned to a very diverse collection of orthodox disciplines: physics, chemistry, biology, geology, meteorology, various branches of engineering, economics, and even small touches of what might be considered psychology and philosophy. Anyone taking a course (or teaching one) based on this book is bound to run into areas where he or she feels uncomfortable and ill-prepared. Courage—charge ahead! Such is the price that must be paid to produce broadly educated scientists and engineers!

This book evolved from a course I taught at the University of Rochester in 1973 and 1974. The students in those classes were the guinea pigs on which my approach was tested, modified, and refined. I wish to thank them for their contribution to this endeavor.

While writing this book, I have had the benefit of conversations with and communications from many individuals, both at the University of Rochester (UR) and elsewhere. Richard Wilson (Harvard University) kindly provided me with a draft of his book in advance of publication. George Hoch (UR) and Robert Knox (UR) struggled to educate me on photosynthesis; both provided valuable criticisms of Chapter 3. Conversations with James C. White (Cornell University) were instrumental in developing the food-chain problems accompanying Chapter 3. Albert Simon (UR) corrected an erroneous view I had concerning fusion (Chapter 5). George Berg (UR) provided me with a detailed critique of Chapter 6 and rewrote portions of that chapter for me. Peter Muhlemann (UR) carefully read the first five chapters and corrected several errors. Robert Collin (UR), Lawrence Lundgren (UR), Laurent Hodges (Iowa State University), Joseph Andrade (University of Utah), and Joseph Priest (Miami University) reviewed the manuscript in its entirety and made many useful suggestions. While all the above-mentioned individuals have provided me with valuable advice, I haven't always followed it. Responsibility for any remaining factual errors and for the philosophical points of view expressed in the book rests solely upon me.

Typing and other technical aspects of manuscript preparation were ably handled by Elizabeth Bauer, who also assisted me with bibliographical matters.

The important connections between energy and environment were first drawn to my attention by my wife, Elizabeth, who suggested that this was an area in the environmental arena where a physicist's talents might be particularly appropriate. It was she who first suggested that the subject matter might make a good college course, and later suggested that the course might make a good book. She found most of the quotes used as chapter openers. Throughout the entire enterprise, she provided encouragement and (when needed) some prodding. She suffered gracefully and cheerfully the considerable household disruption that writing this book entailed. From beginning to end, her role in this project was essential. It is with deep-felt thanks that I dedicate this book to her.

Rochester, New York   E. H. T.
August 1975

Contents

1 Introduction

We cannot command Nature except by obeying her.

Francis Bacon, *Novum Organum*

Energy is the capacity to do work—to exert a force F through a distance d; to accelerate a mass m from rest to some velocity v; to move an electric charge q across a voltage difference V; to lift a mass m to a height h. In each case mentioned, a specific amount of energy (Fd, $\frac{1}{2}mv^2$, qV, or mgh) is required. Since energy is required to make anything "happen," it is not surprising that energy is a central concept in most branches of the natural sciences and engineering. It plays a crucial role in such diverse processes as chemical reactions, cloud formation, functioning of cells, evolution of stars, transmission of radio waves, and behavior of plants and animals in a pond. Energy also plays a crucial role in many human activities such as transportation, production of goods, and home heating. Energy, therefore, should be a central concept in economics and other branches of the social sciences. Until recently, it has not been adequately appreciated by the practitioners of these disciplines, but that is now rapidly changing.

In this book, our concern with energy will focus on how it affects the earth's environment, and how it affects man. We will be concerned with both natural energy processes (e.g., weather and climate, and biological systems) and man-made energy processes (e.g., the generation and distribution of electricity). Natural energy processes exert a major influence upon the natural environment and, therefore, upon man. We shall see that man-made energy processes, in addition to accomplishing their desired results (largely beneficial to man), also have other results which adversely affect the natural and human environments, and are thus harmful to man. Further, we will see that man has been influencing natural energy flows on an increasingly larger scale, and has been doing so not by design, but inadvertently.

Energy comes in many different forms: heat energy (the kinetic energy of random thermal motion of matter on an atomic scale); mechanical energy (the kinetic energy of organized motion of bulk matter); electrical energy (the energy due to the forces that charged bodies exert on one another and that electric currents exert on one another); chemical energy (the energy of chemical bonds, due to electrical forces acting at the atomic level in accord with quantum mechanics); nuclear energy (the energy of nuclear binding due to nuclear forces and coulomb repulsion); gravitational energy (the energy due to the force of gravitational attraction that massive bodies exert on one another—for example, the potential energy that a body of water has because of its height above sea level); light energy (the energy of visible, ultraviolet, and infrared electromagnetic radiation); etc.

Energy undergoes transformations from one form to another. These transformations are governed by the First and Second Laws of Thermodynamics (see Appendix A), which state:

1. Energy is a conserved quantity—while it can move from place to place and change from one form to another, it is neither created nor destroyed. That is, the total amount of energy does not change.

2. In energy transformations, energy tends to pass from concentrated forms to dilute forms. It changes in such a way that the amount of work that can be obtained from it decreases.

As an example of these two laws, consider a very hot, relatively small object placed in contact with the water in a large lake. Heat energy will flow from the hot body to the water in the lake, warming the latter a little and cooling the former a lot. Soon both object and lake will be at the same temperature, slightly above the initial temperature of the lake. The total amount of energy has not changed. Indeed, in this example the total amount of heat energy has not changed. Energy conservation is observed to occur. However, the energy has become much more spread out, more dilute. In the initial arrangement, the hot object could have been used to generate steam to drive a steam engine and do work. The final arrangement—a slightly warmed lake—is much less suitable for obtaining useful work.

Because our interest is in the motion and transformation of energy, we will be concerned with power, energy flow, and energy flux. *Power* has the units of energy per unit time. It is the rate at which work is done, or more generally, the rate at which energy is transformed from one form to another. For example, the power delivered by an electric motor is the rate at which electrical energy is converted to mechanical energy. Power is the time derivative of energy transformed; the energy transformed is the integral of power. *Energy flow* has the same units as power (energy per unit time). It is the rate at which energy is transported—how much energy passes a given point in a given period of time. For example, the amount of energy arriving at the earth's surface from the sun per unit time is the energy flow from the sun to the earth. *Energy flux* is energy

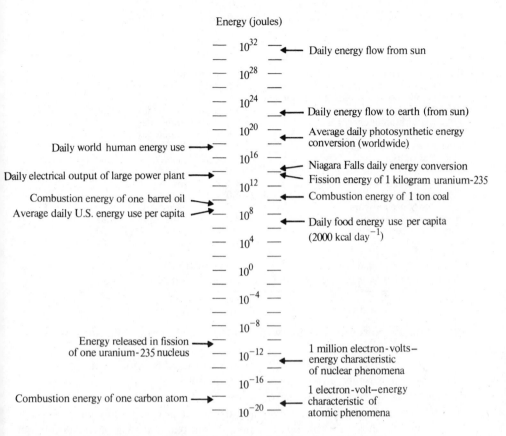

Energy (joules)

10^{32} — ← Daily energy flow from sun

10^{28}

10^{24}

10^{20} — ← Daily energy flow to earth (from sun)

Average daily photosynthetic energy
conversion (worldwide)

Daily world human energy use →

10^{16}

← Niagara Falls daily energy conversion

Daily electrical output of large power plant → ← Fission energy of 1 kilogram uranium-235

10^{12}

Combustion energy of one barrel oil → ← Combustion energy of 1 ton coal

Average daily U.S. energy use per capita → 10^{8}

← Daily food energy use per capita
(2000 kcal day^{-1})

10^{4}

10^{0}

10^{-4}

10^{-8}

Energy released in fission →
of one uranium-235 nucleus 10^{-12} — ← 1 million electron-volts–
energy characteristic
of nuclear phenomena

10^{-16}

1 electron-volt–energy

Combustion energy of one carbon atom → ← characteristic of
10^{-20} — atomic phenomena

Fig. 1.1 Energy of various processes displayed on a logarithmic scale.

flow per unit area. It is the amount of energy that crosses a unit surface constructed perpendicular to the direction of energy flow per unit time. The energy
delivered is the energy flux integrated over the area involved and over time.

Energy transformations and flows take place on many different size scales
from cosmological to subatomic. In our concern for the effects of energy on man
and his environment, we will have dealings with much of this range—from flow
of energy from a star (the sun) at the large end to energy release in chemical
reactions at the small end. The energy and power levels involved cover many
orders of magnitude, as shown in Table 1.1 and Fig. 1.1.

One of the greatest annoyances to be tolerated in any broad treatment of
the subject of energy is the bewildering array of units in which energy is measured. The physicist dealing with macroscopic processes measures energy in
joules and power in watts (joules/sec). The physicist's energy unit for atomic
and subatomic processes is the electron-volt. Biologists and chemists use the
gram-calorie or kilogram-calorie as their energy unit; heating engineers use the
Btu (British thermal unit); electrical energy is sold in kilowatt-hours. Energy is
often quoted in fuel units, i.e., barrels of oil equivalent, tons of coal equivalent,

Table 1.1   Energy, energy flow, and power levels of various processes.

Process	Energy or energy per day	Power or energy flow (watts)
Energy flow from sun	3.2×10^{31} joules/day	3.7×10^{26}
Energy flow to earth (from sun)	1.5×10^{22} joules/day	1.7×10^{17}
Average photosynthetic energy conversion (worldwide)	3.4×10^{18} joules/day	4.0×10^{13}
Total world human energy use	5.4×10^{17} joules/day	6.0×10^{12}
Total U.S. human energy use	1.8×10^{17} joules/day	2.0×10^{12}
Niagara Falls energy conversion	2.7×10^{14} joules/day	3.0×10^9
Electrical output of large power plant	9.0×10^{13} joules/day	10^9
Fission energy of one kilogram uranium-235	8.0×10^{13} joules	—
Combustion energy of one barrel of oil	6.0×10^9 joules	—
U.S. energy use per capita	9.0×10^8 joules/day	10^4
World energy use per capita	2.0×10^8 joules/day	2.0×10^3
Food energy use per capita (2000 kcal/day)	9.0×10^6 joules/day	10^2
Energy released in fission of one uranium-235 nucleus	3.0×10^{-11} joules	—
Combustion energy of one carbon atom	7.0×10^{-19} joules	—

etc. Units for power include the watt, calorie/sec, Btu/hr, and also horsepower. What a mess!

I haven't even done you the favor of writing this book in a single set of units, but have freely hopped from joules to kWh to Btu to barrels of oil, as seemed most appropriate for the occasion. My reasoning is that in the absence of a universally accepted and universally used unit of energy, anyone who reads the energy literature must be able to convert from one unit to another. You might as well accept it now. Appendix E contains definitions of most of the standard units of energy and power, as well as conversion factors between the more common energy units.

In any discussion of human energy use (as with many other human activities), *growth* plays a central role. In this book we will be concerned with the growth of the human population, of human energy use per year, and of various adverse effects of energy use. Growth can follow different patterns. We tend to think of growth as occurring linearly, adding a fixed amount per unit time. For example, the average person gains about three inches in height per year during the first fifteen years. But many processes involve *exponential growth,* in which the amount added per unit time is some fraction of the amount existing at the beginning of the time period. Money in a bank earning compound interest grows in this way, as does the population of a species not limited by food supply, predators, or other factors. People expect their salaries to grow this way (*x* percent more than last year); economists expect GNP to grow this way. Those items of concern to us in this book are growing (roughly) exponentially.

If some quantity I increases by some fraction f of itself per unit time, then in some small time interval dt, the increment dI is given by

$$dI = fI \, dt, \tag{1.1}$$

leading to the differential equation

$$\frac{dI}{dt} = fI,$$ (1.2)

which has the solution

$$I(t) = I_0 e^{ft} = I_0 2^{t/T}; \qquad T = \frac{\ln 2}{f}.$$ (1.3)

In either (equivalent) function, time appears as an exponent, whence the name exponential growth. T has the dimensions of time and is called the *doubling time*, since in the time interval from $t = t_0$ to $t = t_0 + T$, I will double. If f is the fractional increase per year, then the percentage increase per year p is

$$p = 100f,$$

and

$$T = \frac{100 \ln 2}{p} \approx \frac{70}{p} \text{ years.}$$

Thus a quantity that is growing at a rate of 2 percent per year will double in 35 years.

The overriding feature of exponential growth, which is quite evident from the mathematics but for which we are psychologically unprepared, is that annual amounts of growth which are very small initially become extremely large after several doubling times have elapsed. This is illustrated by Problems 1.5 through 1.7.

Questions and Problems

1.1
If 100 joules of energy are used to:
a) accelerate 1 kg of water from rest to a speed v, what is v?
b) lift 1 kg of water to a height h, what is h?
c) raise the temperature of 1 kg of water by an amount ΔT, what is ΔT?
d) evaporate water, what mass of water can be evaporated?
e) decompose water into H_2 and O_2, what mass of water can be decomposed?
(You will need to look up some constants to answer some of these.)

1.2
The average U.S. energy use per unit time is 2.0×10^{12} watts. Express the annual energy use in joules, kWh, Btu, kcal, and eV.

1.3
The energy intake of an inactive adult (basal metabolism) is about 2000 kcal/day. Express this average energy flow into a person (in food), and consequently away from a person (largely as heat) in watts; in horsepower; in Btu/hr.

1.4
A 3-square-meter screen is positioned 8 meters away from a 75-watt light bulb, and oriented with its plane perpendicular to the line between the light bulb and the screen.

What is the energy flux onto the screen? What is the energy flow onto the screen? How much energy strikes the screen in a day?

1.5

A crafty craftsman offered his wealthy monarch an ornately carved chessboard, asking in return only one grain of rice for the first square of the board, 2 grains for the second square, 4 for the third, 8 for the fourth, 16 for the fifth, and so on. The monarch, unfamiliar with the nature of exponential growth, accepted. How many grains of rice were required for the first row of 8 squares? How many *tons* of rice were required for the first half of the board (32 squares)? (There are 50 grains of rice per gram, 10^6 grams/ton.) How many tons were required for the 64th square? (The world annual rice harvest is about 3.0×10^8 tons.)

1.6

A large aquarium has a carrying capacity of 1 million guppies. If the guppy population exceeds that number, crowding, insufficiency of oxygen supply, etc. cause a massive die-off. Until the guppy population reaches 80 percent of the carrying capacity, there are no indications of overpopulation. The aquarium starts off with two guppies, but guppies are prolific, and the population doubles every week. The owner keeps watch on the aquarium with the intention of taking corrective action if there are indications of adverse effects from overpopulation. In how many weeks is the carrying capacity reached? How much advance warning does the aquarium owner get?

1.7

Assume that some planet contains 10^9 tons of a mineral resource (e.g., coal) and the inhabitants of that planet are initially using the resource at a rate of 1000 tons per year. The resource, therefore, constitutes a one-million-year supply at the initial rate of use. Also assume that the rate of use doubles every 20 years. How long will the resource last? Fifty years before the resource is exhausted, how many years' supply remains at the rate of use then prevalent?

Related Reading

Department of the Interior
United States Energy—A Summary Review. Washington, DC: Department of the Interior, January 1972.

The following references are of a general nature and are intended as references for the book in its entirety, not just for Chapter 1.

EIC, Inc.
The Energy Index—A Selected Guide to Energy Literature Since 1970. New York: Environment Information Center, Inc, 1974. Contains a list of energy-related books, including many written before 1970, as well as an index for articles written since 1970, with an abstract of each article.

Scientific American
Energy and Power, A *Scientific American* Book. San Francisco: W. H. Freeman, 1971. This is the September 1971 issue of *Scientific American,* the energy-and-power issue, in book form. It gives a broad treatment of energy, though the "environmental problems" aspect is omitted.

Richard Wilson and William J. Jones
Energy, Ecology, and the Environment. New York: Academic Press, 1974.

2 Global Energy Flows

The play seems out for an almost infinite run.
Don't mind a little thing like the actors fighting.
The only thing I worry about is the sun.
We'll be all right if nothing goes wrong with the
   lighting.

R. Frost, "It Bids Pretty Fair" *

2.0 Introduction

The earth's principal energy source is the sun. Radiant energy from the sun impinges on the earth's atmosphere. Some of this energy is reflected or scattered
into outer space by clouds, by particulate matter, or by gas molecules; some is
absorbed by clouds, particulates, or gas molecules, thus warming the atmosphere;
some penetrates through the atmosphere to the earth's solid or liquid surfaces,
where it is either reflected or absorbed. The reflected energy is subject to all
the above processes as it travels from the earth's surface toward outer space.
The absorbed energy warms the soil or water (becoming *sensible heat*) or
evaporates water (becoming *latent heat*). The sensible heat energy is transferred
to the lower atmosphere by conduction, transported horizontally and vertically
by convection (winds and ocean currents), and transported vertically by reradiation as infrared radiation. The infrared radiation is subject to absorption as it
travels toward outer space. The latent heat travels about in the atmosphere and
is converted to sensible heat when the water vapor condenses. Eventually, all
energy that entered the neighborhood of the earth will leave. Since energy is conserved, and since negligible energy accumulates on the earth, there is a balance

---

* From *The Poetry of Robert Frost* edited by Edward Connery Lathem. Copyright 1928,
1947, © 1969 by Holt, Rinehart and Winston, Inc. Copyright © 1956 by Robert Frost.
Copyright © 1975 by Lesley Frost Ballantine. Reprinted by permission of Holt, Rinehart
and Winston, Publishers.

between energy arriving and energy leaving. (There are short-term deviations from this balance, but averaged over a year, the balance is very good.)

If the preceding gave you the impression of a very complicated situation, then you have the correct picture! All the above-mentioned processes combine to determine the earth's temperature, its climate, and its weather. In this chapter we will use several simplified models to study the physical processes involved. Even our most complicated model will be a far cry from the actual world. (Realistic models currently are being attempted with high-speed computers and require the largest computers available.) Our models will illustrate some of the processes separately—we will concentrate on those involving radiation. Our choice of models is based more on their being soluble than their being realistic. Convection is very important, but will not be modeled because it's too difficult.

2.1   The Sun

The energy flows and processes in the sun are even more complex than those on earth. Nuclear reactions (see Appendix B) take place deep in the sun's interior at extremely high temperatures ($10^7 °K$). Protons combine to form helium in a series of reactions:

$$p + p \rightarrow d + e^+ + \nu$$
$$d + p \rightarrow {}^3He + \gamma$$
$$^3He + {}^3He \rightarrow {}^4He + p + p$$

These reactions are *exothermic* (as opposed to *endothermic*), that is, the final products have less potential energy than the initial reactants, and so energy is released in the form of kinetic energy and gamma rays. (This process, called nuclear fusion, is carried out on earth with hydrogen bombs. Many research projects are searching for methods to carry out nuclear fusion on earth in a non-explosive, controlled fashion, so that useful energy (e.g., in the form of electricity) can be obtained from the process. This topic is discussed in Section 5.3.) The energy released deep in the sun's interior is quickly converted to heat energy, which flows toward the sun's surface by convection and radiation. The temperature of the surface of the sun ($\sim 5800 °K$) is considerably cooler than the interior. The sun is quite opaque, so we see only this relatively cool surface. Energy radiated from the hotter inner regions is reabsorbed before it reaches the surface. From our vantage point on earth, the sun can be treated as a simple black body with a temperature near $5800 °K$.

Energy leaves the sun's surface and travels to earth in the form of electromagnetic radiation, i.e., infrared radiation, visible light, and ultraviolet radiation. The *frequency distribution* of this radiation is to a good approximation that of a black body at $5800 °K$. This means that most of the radiant energy is in the visible and nearby ultraviolet and infrared regions. The frequency distribution (actually wavelength distribution) for a $5800 °K$ black body is shown in Fig. C.2 (see Appendix C).

The amount of solar energy reaching the earth per unit time and per unit area (the *energy flux*) is given by the solar constant S. Imagine a unit area (e.g., one square meter) placed just above the earth's atmosphere and oriented perpendicular to the line joining the centers of the earth and the sun, that is, oriented so that the sun shines squarely on it. Then the amount of energy flowing through this unit area per unit time (e.g., per second) is by definition one solar constant. S has the dimensions of energy/area-time or power/area. It has been measured to be very nearly 1.36 kW/m^2 or 0.136 watts/cm^2.

2.2   The Earth—Our Simplest Model

For our first model of energy flow to and from the earth, we ignore *all* complicating factors. We consider an "earth" with no atmosphere or oceans, and further neglect any variation in temperature with latitude. We assume this model "earth" to be a black body; namely, it reflects *no* incident radiation.

We proceed by requiring an energy balance—the energy flowing to the earth over any long period of time must equal the energy flowing away from it. This must be so, since energy is conserved and since the earth (to a close approximation) is neither accumulating energy nor being depleted of it. The energy flowing to the earth is that coming from the sun. The energy flowing away is the infrared radiation given off by the earth as a warm body, and it can be quantitatively determined from the earth's surface temperature using Stefan's Law.

$$\text{Energy absorbed} = S \times \text{cross-sectional area} \times a \times t = S\pi R^2 at, \qquad (2.1)$$

where a is the absorptivity of the earth's surface, the fraction of the incident radiation that is absorbed. By our assumptions (black body), $a = 1$. R is the radius of the earth and t is the period of time involved.

$$\text{Energy reradiated} = \sigma \epsilon T^4 \times \text{surface area of earth} \times t = \sigma \epsilon T^4 \times 4\pi R^2 t, \qquad (2.2)$$

where σ is the Stefan-Boltzman constant (0.567×10^{-11} joule cm^{-2} deg^{-4} sec^{-1}) and ϵ is the emissivity of the earth's surface. By our assumptions (black body) $\epsilon = 1$. T is the temperature of the earth's surface, assumed to be everywhere the same.

Equating the absorbed and reradiated energy, we have

$$\pi R^2 \, Sat = 4\pi R^2 \, \sigma \epsilon T^4 t, \qquad (2.3)$$

and

$$T^4 = \left(\frac{S}{4\sigma}\right)\left(\frac{a}{\epsilon}\right) = \frac{0.136 \text{ watts cm}^{-2}}{4 \times 0.567 \times 10^{-11} \text{ watts cm}^{-2} \text{ deg}^{-4}}, \qquad (2.4)$$

Therefore,

$$T = 278°\text{K or } 5°\text{C}. \qquad (2.5)$$

Note where the factor 4 crept into the derivation. The area of the earth projected onto a plane perpendicular to the sun's rays is πR^2. This is the area of the disc

Figure 2.1

that absorbs the sun's rays and is the area relevant for absorption (recall the definition of S). On the other hand, all parts of the earth's surface are reradiating, including the side away from the sun. Therefore the earth's total surface area, $4\pi R^2$, appears in the expression for reradiated energy.

Our basic equation in this model is one of energy balance—energy flowing in being equal to energy flowing out. We must investigate what happens when the system becomes slightly out of balance. Does it tend to correct itself or does it run wild? Before doing this, let's consider a much more familiar situation to establish our terminology.

Consider the three marbles shown in Fig. 2.1. All three are in equilibrium, in that in the absence of any disturbance, all three will remain where they are. The marble on the concave surface is in *stable equilibrium*. If we displace it from its equilibrium position it will experience a restoring force, and after some oscillations it will return to the equilibrium position. The initial displacement acts as a cause of a further displacement, a situation referred to as *feedback*. The initial displacement causes the marble to experience a force of sign *opposite* to the initial displacement, giving rise to a displacement back toward the equilibrium position. This situation, with the act tending to oppose itself, is called *negative feedback*. We note that negative feedback gives rise to stable equilibrium.

The marble on the convex surface is in *unstable equilibrium*. If we displace it from its equilibrium position, it will experience a force causing further displacements away from equilibrium, i.e., *positive feedback*. The marble, of course, will roll off the surface. Positive feedback gives rise to unstable equilibrium.

The marble on the flat surface is an example of *neutral equilibrium*. If we displace the marble, it will neither roll away nor return to its initial position. There is *no feedback,* a requirement for neutral equilibrium.

After this lengthy aside, let us consider the stability of our model. Suppose that for some reason the surface temperature of the earth were to drop below the equilibrium temperature. According to Stefan's Law, the rate at which the earth radiates energy outward would also drop. The flux of energy to the earth does not depend on the earth's temperature, so it will *not* drop, and energy will begin to accumulate on the earth, causing it to warm. The chain of events *opposes* the initial temperature drop. The feedback is therefore negative and the system is in stable equilibrium. Similarly, should the temperature fluctuate above equilibrium, the earth will radiate away more energy, cooling itself, and thus return to equilibrium.

Although the model is crude in that it omits all the complicating factors, it does include the basic, fundamental processes—energy flowing to the earth as visible light from the sun warms the earth until the earth reradiates energy as infrared radiation, giving rise to an equal flow of energy away from the earth. The model should yield a numerical result that is "in the ballpark" of the true figure. Further, any more exact model must be a refinement of this one, including the same basic features.

By assuming that the earth's temperature is everywhere the same, we should have arrived at an average temperature for the earth. The observed mean global temperature on the earth's surface is 287°K, or 14°C. Our answer (Eq. 2.5) is too good! With the crudeness of our assumptions, we have no right to be so close! The closeness to the observed value was fortuitous, as will become apparent when we refine the calculation.

2.3 Variation of Temperature with Latitude

It is well known that the temperature is higher near the equator than it is near the poles. We can extend the model used in the previous section and attempt a calculation of this effect. For simplicity, we assume the earth's axis of rotation to be perpendicular to the line from the sun to the earth. In so doing, we have thrown out the seasons. We have also greatly simplified the trigonometry needed to formulate the problem.

Consider a strip on the earth of width W at latitude θ as shown in Fig. 2.2. It will have a circumference of $2\pi(R \cos \theta)$ and an area of $2\pi R \cos \theta W$. This

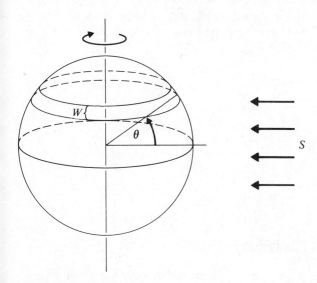

Figure 2.2

is the area relevant to reradiation. If the mean daily temperature at that latitude is $T(\theta)$, then the strip will radiate $\sigma T(\theta)^4\, 2\pi R W \cos\theta$ (power radiated from the strip).

The area of the projection of this strip onto a plane perpendicular to the sun's rays is $(2R \cos\theta) \times (W \cos\theta)$. This is the area relevant to absorption; the strip therefore absorbs $2RW \cos^2\theta\, S$ (power absorbed by the strip).

It is tempting to set these two expressions equal to each other, and we shall do so. However, let us realize what we are assuming in so doing. We are assuming that there is no mechanism that moves energy from one value of θ to another, i.e., toward or away from the equator. This is a much stronger assumption than the one we made in the model in Section 2.2. There we assumed that there were no mechanisms other than absorption and reradiation that carry energy to and away from the earth's surface *as a whole*. Here we are assuming that there are no mechanisms other than absorption and reradiation that carry energy to and away from *a portion* of the earth's surface. This is a very bad assumption, as we shall see.

Equating power absorbed to power reradiated, we have

$$2RW \cos^2\theta\, S = 2\pi RW \cos\theta\, \sigma T(\theta)^4, \tag{2.6}$$

$$T(\theta)^4 = \left(\frac{S}{\sigma}\right)\left(\frac{\cos\theta}{\pi}\right), \tag{2.7}$$

$$T(\theta) = 296\,(\cos\theta)^{1/4}\ {}^\circ\mathrm{K}. \tag{2.8}$$

Values of temperatures calculated with Eq. 2.8 for several different latitudes are listed in Table 2.1. The temperature variation from equator to poles is much too extreme. The calculation fails badly because of our faulty assumption. Both wind and ocean convection currents transport significant amounts of energy from equatorial to polar regions. We cannot expect to obtain a reasonable result for the latitude-dependence of temperature unless convective transport of energy is included in the energy balance equation.

Table 2.1

θ (degrees)	$T(\theta)$
0 (equator)	296°K, 23°C
30	286°K, 13°C
60	247°K, −26°C
84	164°K, −109°C
90 (poles)	0°K, −273°C

2.4 Reflection of Solar Radiation

The model of the preceding two sections assumes that the earth is a black body, namely, that it absorbs all sunlight falling on it. Our everyday experience tells us that this is not the case. Some objects (e.g., clouds, snow fields) reflect a large

Fig. 2.3 Satellite photograph of the earth taken from Synchronous Meteorological Satellite–1 on May 28, 1974. Note the high reflectivity from clouds and from snow in the Andes Mountains. (Photo courtesy of the National Aeronautics and Space Administration.)

fraction of the light striking them. A satellite photograph of the earth (Fig. 2.3) shows this quite clearly. Table 2.2 lists the reflectivity of several surfaces. It is clear that we must modify our model to include reflection of solar radiation.

What we need for our model is a reflectivity averaged appropriately over the entire earth's surface. Specifically, we need to know what fraction of the solar radiation incident on the earth is reflected. This fraction is called the *albedo* and will be denoted by A. It has been measured with satellites orbiting above the earth's atmosphere and has a value near 30 percent. That is, 30 percent of the solar radiation incident on the earth is reflected; the remaining 70 percent is absorbed.

Clouds are the major contributor to the albedo, being responsible for 25 percent. The remaining 5 percent comes mainly from the earth's surface, but also

Table 2.2 Reflectivity of various surfaces.

Object		Reflectivity (percent)
Clouds	Thin cirrus	20
	Altostratus Altocumulus	50
	Cumulus	70
Open water		3–8
Open water, polar regions		25 (due to glancing angle of incidence)
Snow and ice		30–70 (clean, fresh snow is at high end)
Arable land, or Coniferous forest		10–15
Deciduous forest		18
Desert		30

Source: Adapted from SMIC, *Inadvertent Climate Modification*, Report of the Study of Man's Impact on Climate. Cambridge, Mass.: MIT Press, 1971.

from molecules and particles in the atmosphere. Since on the average one-half of the earth's surface is covered by clouds, these clouds must have an average reflectivity near 50 percent in order to contribute 25 percent to the albedo. This is consistent with the data in Table 2.2.

We are now in a position to allow for the fact that the earth is *not* a black body. Consider the mean global temperature as we did in Section 2.2. There we saw that

$$T^4 = \left(\frac{S}{4\sigma}\right)\left(\frac{a}{\epsilon}\right). \tag{2.4'}$$

Treating the earth as a black body, we set $a = 1$, $\epsilon = 1$. Now we wish to set the absorptivity equal to its measured value. Since reflectivity plus absorptivity = 1 for an opaque object such as the earth, we set $a = 1 - A$.

What shall we do with ϵ? It is tempting to claim $a/\epsilon = 1$ because there is a thermodynamic proof to that effect (see Appendix C). However, the correct statement is $a(\nu) = \epsilon(\nu)$, that is, at any given frequency (or wavelength), a and ϵ are equal. What is really meant by the symbols a and ϵ in the equation for T^4 above is the average values of $a(\nu)$ and $\epsilon(\nu)$, *averaged over the relevant range of frequencies.* For our present problem, a (absorptivity) should be averaged over the frequencies important for the *sun's light,* i.e., the visible spectrum; ϵ should be averaged over the frequencies of the *earth's radiation,* i.e., the infrared. Since different frequency ranges are involved (see Fig. C.2), there is no reason that the two need be equal.

If a and ϵ were equal, they would cancel each other. The earth would absorb less but radiate less, and its temperature would not change. However, for snow and clouds, visible light is easily reflected ($a \ll 1$), whereas these objects are black to infrared radiation, absorbing it almost completely. In fact, it is a reasonable approximation to treat the earth as *black* to *infrared* radiation, i.e., to take

$\epsilon = 1$. As discussed at the beginning of this section, it is not a good approximation to treat the earth as black to visible light.

We therefore set $\epsilon = 1$, $a = 1 - A$, so that

$$T = \left(\frac{S(1 - A)}{4\sigma}\right)^{\frac{1}{4}}. \tag{2.9}$$

Note that the only difference between this result and that for the "black-body earth" of Section 2.2 is that the solar constant S is multiplied by the factor $1 - A$. That is, the temperature is determined by only that fraction of the solar constant that is actually absorbed. Using the measured value for A, 0.3, we obtain an average temperature $T = 255°K$, or $-18°C$. This result is worse than our first, crude result. We'll see why in the next few sections.

As seen above, the albedo influences the mean global temperature. But the mean global temperature influences the amount of snow cover, a contributor to the albedo. Such a system involves feedback, and we must look into the question of stability.

Consider a slight drop in temperature below equilibrium. This will lead to less melting of snowfields, a more extensive snow cover, and a larger albedo. But the larger albedo means a lower equilibrium temperature, and so the feedback is *positive* and the system *unstable*. Indeed there are serious concerns that initially small temperature changes will be amplified by this positive feedback mechanism, resulting in major climatic changes. For example, a slight cooling of the earth near the north pole may cause the northern glaciers and ice fields to grow, reflecting back more sunlight. This lowers the temperature there further, etc., etc., leading to a new ice age. Alternatively, a slight warming of the earth near the poles may cause the snow and ice fields to recede, reflecting less sunlight. This raises the temperature there further, etc., etc., leading to complete melting of the polar ice caps. The crucial point is that the system appears unstable; the feedback is positive.

2.5   The Atmosphere—Radiative Effects

The earth's atmosphere as well as the clouds and particles in it can reflect, scatter, absorb, and transmit radiation. This applies to the incident solar radiation, solar radiation reflected from the earth's surface, infrared radiation from the earth's surface, and infrared radiation from other parts of the atmosphere. A detailed consideration of everything that is going on would fill an entire book. Here we will consider the basic features of two of the more important processes—absorption of ultraviolet radiation and absorption of infrared radiation. We have already discussed a third atmospheric process—reflection of solar radiation—in Section 2.4.

Radiation at wavelengths shorter than 0.30μ is dangerous to humans (sunburn, etc.) and to life in general. Photons at these short wavelengths are sufficiently energetic to readily break the bonds in molecules of living matter. Fortunately it is filtered out (absorbed) by the atmosphere and very little reaches the earth's surface. Wavelengths shorter than 0.18μ are strongly absorbed by molecular oxygen and are therefore removed before the sunlight has penetrated very deeply into the earth's atmosphere. This typically takes place in the ionosphere at an elevation of 100 kilometers. The wavelength region between 0.18μ and 0.29μ is less strongly absorbed, and so can penetrate deeper into the atmosphere through the ionosphere and mesosphere. As the radiation penetrates lower into the atmosphere, it passes through regions of increasingly higher density. Eventually it too is absorbed, mainly in the stratosphere (15 to 50 km), by molecular oxygen and ozone. Wavelengths longer than 0.29μ are only weakly absorbed, mainly by water vapor, clouds, and dust, in the lower troposphere. Approximately 20 percent of the incident radiation in this wavelength interval is absorbed before reaching the earth's surface. From the point of view of the earth's energy balance, this is the important wavelength range, since it contains more than 97 percent of the incident solar energy.

Though unimportant in considering the overall energy balance, the narrow region of wavelengths between 0.295μ and 0.315μ is worth special attention, since it is dangerous to humans and only marginally absorbed by the atmopshere. This radiation is partially absorbed in the stratosphere by ozone. Changes in the ozone concentration in the stratosphere (conceivably caused by activities of man, such as flights of the supersonic transport) would change the amount of ultraviolet radiation reaching the earth's surface.

While the atmosphere is largely transparent to the incident solar radiation (in the visible and near ultraviolet and infrared regions), it is rather opaque to the earth's reradiation (in the infrared region 5μ to 50μ). This is due to the presence of carbon dioxide, water vapor, and clouds. Absorption coefficients for CO_2 and H_2O are shown in Fig. 2.4. CO_2 absorbs IR radiation between 12μ and 18μ. H_2O absorbs IR radiation between 5μ and 8μ and beyond 19μ. Clouds are essentially black to IR radiation.

Because of the above, heat energy radiated from the earth's surface will not disperse directly into outer space, but will be absorbed by the atmosphere. The piece of atmosphere absorbing radiation from the earth will be warmed and will reradiate the energy. However, it will reradiate isotropically, namely, half toward outer space and half back toward the earth. This energy radiated back toward the earth will warm the earth's surface, maintaining it at a *higher temperature* than would be the case if the atmosphere were transparent to infrared. This is the so-called *greenhouse effect*.

We can illustrate the greenhouse effect with a simple model. Consider the earth with an albedo of 0.3 (as in the model of Section 2.4), but now surrounded by a shell (e.g., of CO_2) that is transparent to the sun's energy but opaque to infrared. Determine the temperature of the shell and of the earth's surface.

Since the shell is transparent to the sun's energy, the energy flowing in is the same as it would be if the shell were not there. For an energy balance, the energy flowing out must also be the same. But now the energy flowing out must be emitted from the outer surface of the shell, since all energy radiated by the earth's surface is stopped by the shell. It follows that the shell will have the temperature

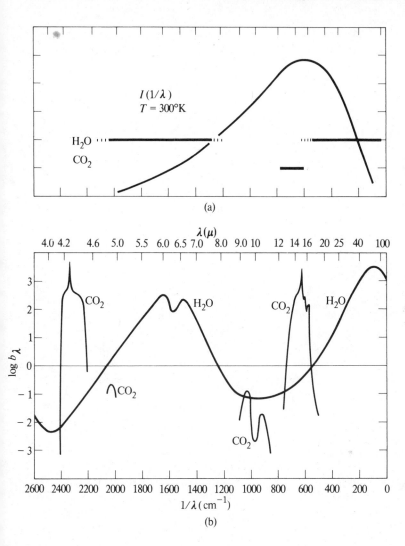

Fig. 2.4 Infrared absorption by CO_2 and H_2O. (a) The intensity distribution characteristic of the earth's infrared radiation, with CO_2 and H_2O absorption regions indicated. (b) The natural logarithm of the absorption parameter b_λ plotted against $1/\lambda$. The approximate transmission is given by $t = 2^{-(b_\lambda x/x_0)}$, where x is the amount of material (CO_2 or H_2O) to be traversed, and x_0 is the amount traversed in passing entirely through the atmosphere under "normal" conditions. For H_2O, $x_0 = 1$ g/cm²; for CO_2, $x_0 = 0.43$ g/cm². For values of $\log b_\lambda$ much below zero, atmospheric absorption is small. (From C. W. Allen, *Astrophysical Quantities,* 3rd edition. New York: The Athlone Press, 1963, p. 126. By permission.)

T of 255°K that we previously calculated for the earth's surface without a shell. Now, not only will the outer surface of the shell radiate energy to outer space,

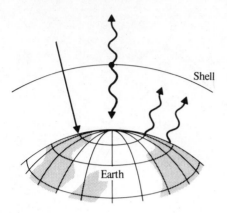

Figure 2.5

but also the inner surface of the shell will radiate energy back to the earth's surface. Assuming the outer and inner surfaces of the shell are at the same temperature, the shell will radiate equal amounts of energy outward and inward. The earth's surface will receive twice as much energy as it would if the shell were not there—one unit from the sun; the other unit from the shell. Its temperature will therefore rise until it radiates out twice as much energy. The energy flows are indicated in Fig. 2.5. You can verify that the energy flows to and from each object (shell and earth) balance. The temperature of the earth's surface necessary for it to radiate twice as much energy as in the "no-shell" case (Section 2.4) is

$$T = 255°K \times 2^{1/4} = 304°K.$$

This answer is somewhat higher than the actual mean global surface temperature.

Our simple "shell model" shows how the name "greenhouse effect" originated. The shell plays the role of the glass roof, which transmits sunlight but absorbs infrared radiation, reradiating it half up, half down. (In addition, the glass over a greenhouse reduces convective losses.)

It is fairly easy to generalize the above model to the cases where the earth is surrounded by two, three, or any arbitrary number of concentric shells (see Problem 2.6). The result is that the more shells there are, the higher the earth's surface temperature. The temperature of the outermost shell does not change, however, staying at 255°K. The temperatures of the shells in between take on values intermediate between the earth's surface and the outermost shell, decreasing monotonically as one moves outward.

A more realistic model assumes that the IR-absorbing material (CO_2, H_2O, clouds) is distributed vertically in a continuous fashion rather than in discrete shells. This model requires more sophistication to formulate and solve, but is still quite tractable (see Problem 2.7). The conclusions that follow are qualitatively the same.

1. The more IR-absorbing material, the higher the earth's surface temperature.

2. This type of model, allowing for the radiative effects of the atmosphere (albedo, greenhouse effect) but omitting convection, gives too high a value for the earth's surface temperature.

3. The models predict that the temperature of the atmosphere will fall with increasing altitude. This is observed to be the case from the earth's surface up through the troposphere to the tropopause, which is the boundary between the troposphere and the stratosphere. (In the stratosphere, the temperature rises with increasing altitude, due to strong absorption of UV in the upper stratosphere.) However, the fall in temperature with increasing altitude through the troposphere is actually less than that predicted by the model.

2.6   Convection and Evaporation

So far the only thing we have allowed to *move* has been electromagnetic radiation. We have gone as far as we can along that road and now must admit that energy moves around in other ways.

Evaporation is a major "use" of solar energy. Of the incident solar energy, 30 percent is reflected, 47 percent is converted directly to sensible heat (i.e., warms things), and 23 percent is converted to latent heat (evaporation). As the water vapor moves with the winds, the latent heat is transported with it, to become sensible heat when the vapor condenses.

There are substantial amounts of latent and sensible heat transferred horizontally by *convection*. The major winds are driven by the temperature differences between equator and poles. The winds, in turn, drive the ocean currents. The situation is complicated by the Coriolis force, by the land masses, and by seasonal changes. The net effect is that both wind and ocean currents carry comparable amounts of energy from equatorial regions to polar regions. This reduces the temperature differences between equator and poles below that calculated for radiative effects alone, and therefore explains the poor answer obtained in Section 2.3.

Convection also transports energy vertically. The temperature drop from the earth's surface to the upper troposphere (caused by the greenhouse effect) sets up vertical air currents, which carry latent and sensible heat upward. This reduces the temperature difference below that indicated by radiative effects alone.

Although they are very important, we will attempt no models of these processes. Models must either be solved numerically or must use empirical information as input. In other words, modeling is too messy, so we will skip it. (See, however, Problem 2.9.)

To summarize this section, the temperature differences created by radiative effects are reduced by convective effects. The process yields winds, ocean currents, transport of water vapor from ocean to land, rainfall, etc., namely, weather and climate.

2.7   Human Influence on Global Energy Flows

Man has already modified *local* climates. Specifically, the climate in cities differs from the surrounding countryside. (Cities are a degree or two warmer, and somewhat rainier.) At first thought, it would seem unlikely that man could influence the large-scale energy flows that determine continental and global climates. The power levels involved in nature dwarf those of man (see Table 2.3).

Table 2.3

Mode of energy flow	Power level
Solar radiation	$173,000 \times 10^{12}$ watts
Direct reflection	$52,000 \times 10^{12}$ watts
Direct conversion to heat	$81,000 \times 10^{12}$ watts
Evaporation	$40,000 \times 10^{12}$ watts
Man's power use (1970)	6×10^{12} watts

However, two considerations suggest we give the matter more careful thought. First, it seems probable that man *already has* modified the climate of fairly large areas. It is estimated that 20 percent of the land surface of the earth has been drastically altered by man, much of this alteration occurring in pre-industrial times. Forests have been converted to grasslands; overgrazing has converted grasslands to semideserts. Large-scale surface modifications change the reflectivity and therefore the radiative balance. Reduction in vegetation also alters the distribution of energy between latent and sensible heat. The result is that the climate of a region can be permanently changed. Specifically, the rainfall of the area may drop. There is evidence that some of the world's deserts are man-made. Note that these changes were caused by civilizations using far less power than we do today.

Second, the reasoning comparing the power levels of man and nature assumes that change must be caused by direct assault. But legend is full of cases of the weak but clever causing the strong to trip over their own strength. The question is one of stability. Global climate results from a delicate balance among very large energy flows. A small imbalance could lead to considerable climatic changes. If global climate is a stable system, then considerable energy will be required to change it. Alternatively, if it is an unstable system, some small change may serve as a trigger to "trip" it.

We have already seen one positive feedback mechanism influencing climate, namely, the snow-cover/arctic-ice/albedo loop. Another is the water-vapor/greenhouse coupling. A slight rise in temperature will increase the amount of water vapor in the atmosphere. This in turn absorbs a larger amount of infrared radiation, causing an increase in the greenhouse effect and a further rise in temperature. With these positive feedback mechanisms present, the possibility exists that a small initial action could result in a large change. Here are some processes observable today that are potential trigger mechanisms.

1. Increased atmospheric CO_2 from burning fossil fuels leads to an enhanced greenhouse effect and a temperature rise.
2. Increased cloudiness caused by air pollution (condensation nuclei) causes albedo to go up and T to go down. But cloudiness also makes the greenhouse effect go up, causing T to rise. The net effect might go either way.
3. Waste heat from man's use of power (if it continues to grow at its present rate) causes a temperature rise.
4. Land surface alterations cause a change in surface reflectivity and in the distribution of energy between sensible and latent heat.

Some of these mechanisms will be considered in greater detail in Chapter 6. Here we only stress that man's inadvertent modification of climate is quite within the realm of possibility. It could have very serious consequences.

Questions and Problems

2.1
Give several examples of systems with negative feedback. Give some examples of systems with positive feedback. Wherever possible, write down and solve the differential equation describing the system.

2.2 *
Consider quantitatively the stability of the simple model of the earth's energy balance given in Section 2.2. Specifically, assume that at time $t = 0$, the earth's temperature is T_1 rather than the equilibrium temperature T_0.

a) Show that the earth's temperature T will change from T_1 to T_0 according to the equation

$$T - T_0 = (T_1 - T_0)e^{-t/\tau},$$

assuming $T_1 - T_0 \ll T_0$. Show that $\tau = C_A/4\sigma T_0^3$, where C_A is the heat capacity per unit area of the earth.

b) Assume there are contributions to C_A from the atmosphere (1 kg air/cm²), the top 50 meters of the ocean (covering 70 percent of the earth's surface), and the upper 10 cm of the solid earth (30 percent of the earth's surface). Evaluate C_A and τ.

2.3 †
a) Calculate the mean daily radiation intensity *vs.* time of year for several different latitudes, allowing for the fact that the earth's axis of rotation is tilted 23 deg away from the perpendicular to the earth's plane of revolution about the sun. Present your results graphically. (*Warning:* The situation is hard to visualize and the trigonometry is rather messy. You may find it helpful to play with a globe (for visualizing), and to use a computer to integrate over a 24-hour period rather than obtain an analytic expression.)

---

* Difficult. This designation will follow throughout the Question and Problems in this book.
† Very difficult. This designation will follow throughout the Questions and Problems in this book.

b) Convert the mean daily radiation intensities to mean daily temperatures, making the (very bad) assumptions of (1) no movement of energy from the equator toward the poles (as in Section 2.3), and (2) negligible heat capacity for the earth, so that the equilibrium temperature is reached instantly. Plot calculated mean daily temperature *vs.* time of year for three latitudes: (1) the equator, (2) the north pole, (3) ~45 deg latitude. Also plot observed mean daily temperatures (obtained from Weather Bureau records).

c) Compare calculation and observation, giving attention to the *magnitude* and *phase* of the seasonal temperature variation. Explain the observed discrepancies in both magnitude and phase.

2.4

Suppose the earth's average cloud cover were to increase from 50 percent to 52 percent due to particulate air pollution. Using a simple model, calculate the change in the earth's mean surface temperature resulting from the increase in albedo.

Discuss the validity of your model, bearing in mind that you are using it only to determine a temperature *change*. Mention any important processes that have been neglected. How would you expect the actual change in mean surface temperature resulting from a 2-percent increase in cloudiness to compare with your calculated value?

2.5 *

Assume the average cloud cover of the earth were proportional to the earth's surface temperature. (This is a bad assumption, made here only for purposes of this problem.) Assume further that the average reflectivity of clouds is 50 percent (a good value), and that the reflectivity of the earth's surface is zero (a bad value, given to simplify the problem). If the relation between cloud cover and temperature were

$$\text{Cloud cover} = 0.50 \times \frac{T}{300°\text{K}},$$

what temperature would the earth's surface assume? Discuss stability, feedback. (In this problem, ignore all aspects of clouds and atmosphere except those explicitly mentioned, i.e., ignore convection and IR absorption.)

2.6

a) Consider an "earth" which is a black body surrounded by two thin, concentric shells of CO_2. Each shell is transparent to visible radiation but opaque to infrared. Calculate the average temperature of the earth's surface and of each CO_2 shell.

b) Repeat the problem for the case of three concentric shells.

c) Generalize to n concentric shells.

2.7 * ‡

Consider a black-body "earth" with a stationary atmosphere (no convection) of depth L and uniform density. Assume the atmosphere is transparent to visible light but attenuates infrared radiation. Let the fraction attenuated by a short path length Δx be $\Delta x / \lambda$, where λ is a constant.

a) Obtain differential equations governing $T(z)$, the equilibrium temperature a distance z above the earth's surface, $0 \lesssim z \lesssim L$. (*Hint:* Consider an infinitesimal layer of

---

‡ Needs advanced physics or math. This designation will follow throughout the Questions and Problems in this book.

atmosphere of thickness Δz. Introduce variables $IU(z)$ and $ID(z)$ for the upward-moving and downward-moving infrared fluxes, and consider how they change in crossing this layer. Additionally, require a balance between radiation absorbed and emitted by the layer. The emissivity of the layer equals its absorbtivity, $\Delta z/\lambda$.)

b) Specify the appropriate boundary conditions at $z = 0$ and $z = L$. Solve the differential equations, obtaining $T(z)$.

c) Compare your solution in (b) with the answer in Problem 2.6 above. Relate n, the number of shells in that problem, to L/λ, the number of attenuation lengths in the atmosphere.

2.8
Convection in the earth's atmosphere results because warm air is lighter than cool air. Imagine a gas for which the opposite were true. How would convection function in that case? Describe the heat flow on a planet with an atmosphere composed of such a gas.

2.9 * ‡
In Section 2.6, no model was attempted for convective or evaporative energy transfers because a "first principles" approach was too complex. However, it was noted that the effect of these processes was to transfer energy from high-temperature regions to low-temperature regions. Since conduction also transfers heat energy in this fashion, one might consider a *phenomenological* model, in which evaporative and convective energy transfers are replaced by conductive transfers. In such a model, the thermal conductivity used would be far above the actual physical value, and would be chosen so as to obtain agreement with the observed temperature distributions of energy transfers.

a) In this vein, attempt a model of the earth that gives the variation of temperature with latitude, by augmenting the approach of Section 2.3 with the assumption that the earth has an effective surface thermal conductivity κ. Plot $T(\theta)$ *vs.* θ for several values of κ. (*Hint:* The differential equation obtained is probably best solved numerically, by an iterative technique, using a computer.)

b) If you've been successful so far and are very ambitious, try to obtain the seasonal temperature variation in this way. Extend the model by tilting the earth's axis through 23 degrees (see Problem 2.3) and allowing for a surface heat capacity per unit area C_A (see Problem 2.2). Obtain the partial differential equation satisfied by $T(\theta,t)$. Solve in whatever way you can. Adjust κ and C_A to obtain the best agreement with the actual mean daily temperature as a function of latitude and time of year.

Related Reading

Theodore L. Brown
Energy and the Environment. Columbus, Ohio: Charles E. Merrill, 1971. The first half of this book gives an elementary, qualitative treatment of the earth's large-scale energy flows, with particular attention to radiative phenomena in the atmosphere. Aimed at the lay reader.

SMIC
Inadvertent Climate Modification, Report of the Study of Man's Impact on Climate. Cambridge, Mass.: MIT Press, 1971. Contains chapters on the normal functioning of climate (Chapter 5) and climate modeling (Chapter 6).

3 Biological Energy and Ecosystems

Our life runs down in sending up the clock.
The brook runs down in sending up our life.
The sun runs down in sending up the brook.
And there is something sending up the sun.

R. Frost, "West-Running Brook" *

3.0 Introduction

We had better start by defining the term *ecosystem*. It refers to all the living and nonliving things in some specified area such as a pond, a field, or a forest. In studying an ecosystem, the concern is with the *interactions between the various parts*. And one aspect of ecosystems that has received considerable attention from ecologists is the way that energy flows through the systems.

The vast majority of the energy in an ecosystem flows purely by physical means. For example, sunlight warms a pond's surface, evaporating water. We have covered such processes in Chapter 2 and will not consider them further here. In this chapter our attention will be on *biological* energy flows. Since photosynthesis is the crucial energy link between the physical and the biological worlds, we shall first consider it in some detail. We shall then consider how energy flows from one living member of an ecosystem to another. Finally, we will compare man's scheme for obtaining biological energy—agriculture—with the processes in a natural setting.

Physical processes are invariably observed to "run downhill" . . . to go from

---

* From *The Poetry of Robert Frost* edited by Edward Connery Lathem. Copyright 1928, 1947, © 1969 by Holt, Rinehart and Winston, Inc. Copyright © 1956 by Robert Frost. Copyright © 1975 by Lesley Frost Ballantine. Reprinted by permission of Holt, Rinehart and Winston, Publishers.

333.7   T394e

c.1

states of concentrated energy to states of diffuse energy . . . to go from order to disorder. This observation is quantified by the Second Law of Thermodynamics with all its many statements (see Appendix A). Biological processes generally, and photosynthesis in particular, appear to "go uphill" . . . to violate the Second Law. Of course, the Second Law is not actually violated. The physical components of a process must be considered along with the biological. The process as a whole, biological plus physical aspects, still runs downhill, entropy still increases, etc. Viewing the biological process as an engine trying to extract work from the accompanying physical processes, we observe that its efficiency is always below the thermodynamic limit specified by the Second Law. Throughout this chapter, we will focus on the *efficiency* of energy transfer . . . what fraction is used and what fraction is wasted.

3.1 Photosynthesis—A Macroscopic View

Photosynthesis is the crucial link by which energy enters biological systems from the physical world. Its importance cannot be overstated. Technically, it's not the only link—when you are warmed by the sun, you are picking up heat energy. However, photosynthesis is the only link supplying *storable* energy. All the motive power of living things, both plants and animals, and all their stored, chemical energy comes ultimately from the sun through photosynthesis. Photosynthesis is carried out by green plants and also by some species of bacteria.

In its simplest form, the photosynthetic reaction may be written as:

$$H_2O + CO_2 + light \rightarrow CH_2O + O_2, \tag{3.1}$$

or more generally, water + carbon dioxide + light $\rightarrow$ carbohydrate + oxygen. This reaction "accomplishes" two things. (1) Carbon is "fixed," that is, converted from the inorganic form CO_2 to the organic form $C_nH_{2n}O_n$, namely, carbohydrate. This latter form is a basic constituent of living matter, and the carbon source for other forms such as proteins and fats. Photosynthesis, therefore, provides the "building material" for living organisms. (2) The energy of the sunlight is stored, that is, converted to concentrated chemical energy. It is then available as the energy source for living matter, being "liberated" by the reaction going in reverse: carbohydrate + oxygen $\rightarrow$ carbon dioxide + water + energy.

Let's look quantitatively at the energetics. Consider burning glucose, a typical carbohydrate:

$$C_6H_{12}O_6 + 6O_2 \rightarrow 6CO_2 + 6H_2O + 690 \text{ kcal/mole glucose} \tag{3.2}$$

From this we know the energy that must be *added* for the inverse reaction, namely, 690 kcal/mole of glucose produced. Let's convert from moles to molecules, and from kcal to electron-volts (eV).

$$\frac{\text{Energy}}{\text{Glucose molecule}} = \frac{690 \text{ kcal/mole} \times 4.2 \times 10^3 \text{ joule/kcal} \times 0.63 \times 10^{19} \text{ eV/joule}}{6 \times 10^{23} \text{ molecules/mole}}$$

$$= 30 \text{ eV/glucose molecule.} \tag{3.3}$$

This implies that 5 electron-volts of energy are required for each carbon atom that is fixed.

Now recall that light is made up of "particles" called photons, and that the energy of a photon is proportional to the frequency of the light: $E = h\nu$. We pose the question: If only one photon were required to fix one carbon atom, what frequency and wavelength would it have? Using $h = 4.2 \times 10^{-15}$ eV-sec, we find $E \gtrsim 5$ eV implies $\nu \gtrsim 1.2 \times 10^{15}$ Hz, $\lambda \lesssim 2500$Å. This wavelength is well into the ultraviolet range and is completely absorbed high in the atmosphere. Therefore, it must require more than one photon to fix each carbon atom. Why? Is there some "advantage" to this more complicated approach?

Recall the discussion of atmospheric absorption of UV in Section 2.5, where it was pointed out that wavelengths shorter than 3000Å were hazardous to life. There is an energetic basis for this figure. The basic component of the molecular structure of living matter is the carbon chain, as exemplified by glucose:

$$
\begin{array}{c}
\text{H} \quad \text{H} \quad \text{H} \quad \text{H} \quad \text{H} \quad \text{H} \\
| \quad\ | \quad\ | \quad\ | \quad\ | \quad\ | \\
\text{H--C--C--C--C--C--C=O} \\
| \quad\ | \quad\ | \quad\ | \quad\ | \\
\text{OH OH OH OH OH}
\end{array}
$$

The single carbon bond C—C is necessary for living matter to exist. The minimum energy required to break that bond is 3.6 eV, corresponding to a photon of 3500Å. If light were present with substantial intensity at wavelengths much shorter than 3500Å (energies much above 3.6 eV), then the molecules of living matter would be constantly breaking. Nature is given the task of supplying 5 eV worth of solar energy for fixing a carbon atom in photosynthesis, without endangering the 3.6-eV bond of the carbon chain. Using several low-energy photons for each carbon atom fixed is a natural solution.

The experimental observation is that eight photons are required for each carbon atom fixed. Their wavelength must be 7000Å or shorter. A 7000Å photon has an energy of 1.8 eV, so that the maximum efficiency for photosynthesis (using the lowest possible energy input) is:

$$
\frac{5 \text{ eV}}{8 \times 1.8 \text{ eV}} \approx 35\%. \tag{3.4}
$$

How does this compare with the maximum possible efficiency as given by the Second Law?

One can think of photosynthesis as a heat engine operating between the solar radiation and the green plant. The maximum possible efficiency is $1 - T_P/T_R$. T_P is the temperature of the plant, about 300°K. T_R is the temperature of the solar radiation. At first, one might think this should be the temperature of the surface of the sun (5800°K), yielding a maximum possible efficiency of 95 percent. By that standard photosynthesis seems inefficient. However, it pays to reflect a bit more on what is meant by the "temperature of the solar radiation." If the sun's light comes to a point on earth directly, unscattered by clouds, etc., then by a system of lenses or mirrors one can focus that light onto a very small region and obtain a very high temperature there, in principle as high as 5800°K (but no higher, in keeping with the Second Law of Thermodynamics). But such a focusing system would be useless for the diffuse radiation reflected from clouds, dust in the sky, etc. It would work only if it could track the sun. A system designed for collecting the diffuse radiation cannot achieve as high a temperature, but will still work without tracking the sun, when the sun is out of the direct line-of-sight, on cloudy days, etc. It appears that solar radiation still capable of being focused has a temperature of 5800°K. But diffuse, scattered solar radiation has a much lower temperature.

Speaking teleologically,[1] nature had a choice. To store energy, nature could choose either a system which would take advantage of the high temperature of direct solar radiation but which would fail if it could not track the sun, or alternatively a system which would work on diffuse solar radiation, at lower efficiency but for a larger fraction of time. The latter system would work on direct solar radiation but would not take advantage of its higher temperature. Nature chose the latter system in designing photosynthesis.

Can we estimate the temperature of diffuse solar radiation? Yes, by resorting to the concept of *entropy*. (Those not already familiar with the concept might as well skip on to the answer.) The temperature of a radiation field can be defined as the ratio of the energy of the field to the entropy of the field:

$$T_R = \frac{E}{S}.$$

To determine the temperature reduction in a radiation field because it is diffuse rather than direct from a small region, we need only determine the entropy increase. (We are considering a radiation field whose energy E does not change, but whose entropy changes because the photons are now going in all directions rather than all being nearly parallel.) From its statistical mechanical definition, entropy is proportional to the log of the thermodynamic probability W:

$$S = k \ln W,$$

where the constant of proportionality k is Boltzmann's constant. Let W_0 be the thermodynamic probability for the radiation from a small solid angle Ω (i.e., directly from the sun), and W_1 refer to isotropic radiation coming from a solid angle 4π. Then

$$W_1 = \frac{4\pi}{\Omega} W_0 \tag{3.5}$$

$$S_0 = \frac{h\nu}{T_0} = k \ln W_0 \tag{3.6}$$

$$S_1 = \frac{h\nu}{T_1} = k \ln W_1 = k \ln W_0 + k \ln \frac{4\pi}{\Omega}, \tag{3.7}$$

$$\therefore \frac{h\nu}{T_1} = \frac{h\nu}{T_0} + k \ln \frac{4\pi}{\Omega} \tag{3.8}$$

$$T_1 = T_0 \left(\frac{1}{1 + \dfrac{kT_0}{h\nu} \ln \dfrac{4\pi}{\Omega}} \right). \tag{3.9}$$

The sun's radius is 0.7×10^6 km, and the earth-sun separation is 149×10^6 km. Therefore, the light from the sun's surface to a point on the earth comes from a solid angle

$$\Omega = \frac{\pi(0.7 \times 10^6 \text{ km})^2}{(149 \times 10^6 \text{ km})^2} \approx \frac{\pi}{4} \times 10^{-4}. \tag{3.10}$$

$T_0 = 5800°\text{K}$, and the photons of interest are those with energy just able to take part in photosynthetic processes, i.e., $h\nu \approx 1.8$ eV. Evaluating, we find

$$T_1 \approx 5800°\text{K} \left(\frac{1}{1 + \dfrac{0.5 \text{ eV}}{1.8 \text{ eV}} \ln 1.6 \times 10^5} \right) \approx 1340°\text{K}. \tag{3.11}$$

This temperature of 1340°K is appropriate for isotropic red solar radiation. But the photosynthetic process works at light intensities lower than this, corresponding to lower radiation temperatures (i.e., a "cooler sun"). By determining the minimum light intensity necessary for efficient photosynthesis, and by calculating the temperature corresponding to *that* radiation field, one finds the minimum radiation temperature at which photosynthesis works well. It turns out to be about 1100°K. This appears to be the "design temperature" for the photosynthetic engine; it will work from radiation fields of higher temperature, but its *efficiency* will not rise with the temperature increase.

Summarizing, the photosynthetic process is designed to work on low-intensity, isotropic radiation, corresponding to a temperature of 1100°K. This gives it a maximum thermodynamic efficiency of

$$1 - \frac{300°\text{K}}{1100°\text{K}} = 73\%.$$

Had the photosynthetic process been designed to work from a higher temperature radiation field, a higher maximum thermodynamic efficiency would have been possible. However, the higher temperature radiation field is not always available (cloudy days). Nature has sacrificed efficiency for dependability.[1]

To understand the discrepancy between the maximum possible efficiency of 73 percent and the observed efficiency of 35 percent, we must take a look at the molecular chemistry of the photosynthetic process.

3.2   Photosynthesis — A Molecular View

At the molecular level, photosynthesis consists of the transfer of hydrogen atoms from water molecules to carbon dioxide molecules, using energy from absorbed photons.

Before	$2H_2O$	CO_2
Four hydrogen atoms transferred		
After	O_2	$CH_2O + H_2O$

The process can be divided into three stages:

1. Light collection,
2. Oxidation of H_2O to O_2, with the formation of strong reducing molecules (NADPH) and energy-rich molecules (ATP),[2] using the energy collected from the light,
3. Reduction of CO_2 to CH_2O and H_2O, using NADPH and ATP.

Light Collection

Light falling upon a photosynthetic system, for example, the leaf of a tree, is absorbed by the chlorophyll and other pigments, exciting the pigment molecules (i.e., raising them to a state of higher energy). The reaction may be written

$$\gamma + M \rightarrow M^*,$$

where γ indicates the photon, M the unexcited molecule, and M^* the excited molecule. Chlorophyll absorbs light strongly in two bands, one near 6700Å and the other near 4300Å. This indicates that the chlorophyll molecule has excited states with energies 1.88 eV and 2.9 eV above the ground state. A blue photon will excite the 2.9-eV state, whereas a red photon will excite the 1.88-eV state. When the 2.9-eV state is excited, it will usually decay to the 1.88-eV state, "wasting" 1.02 eV of energy. A molecule excited to the 1.88-eV state can transfer its energy to a nearby unexcited molecule $(M_1^* + M_2 \rightarrow M_1 + M_2^*)$. In this way the energy travels from molecule to molecule until it is transferred to a special type of chlorophyll molecule that has an excitation energy slightly *lower* than 1.88 eV. A small amount of energy is "wasted" (i.e., converted to heat) in the transfer from standard chlorophyll molecule to "trap" molecule. The remaining energy is now trapped—it cannot be transferred to a standard chlorophyll molecule because there is insufficient energy to excite it. The small amount of wasted energy is the "price to be paid" for the service of trapping the energy.

In green plants, the basic photosynthetic unit consists of about 300 chlorophyll molecules and one trap molecule. The function of the 300 chlorophyll molecules is to collect and "deliver" the light energy to the trap. The chemical reactions take place at the trap. The photosynthetic unit contains other pigment molecules in addition to chlorophyll (e.g., carotenoids). These molecules have energy levels *higher* than 1.88 eV and absorb light at wavelengths which the chlorophyll absorbs poorly, i.e., those between red and blue. Energy can be transferred from an excited carotene molecule to a chlorophyll molecule, thereby exciting it $(M_{CA}^* + M_{CL} \rightarrow M_{CA} + M_{CL}^*)$. Note that in this reaction also, energy is "wasted." The excitation energy of chlorophyll is less than that of carotene. Because of the energy-level difference, this reaction does not proceed in reverse to any appreciable extent. The function served by the carotene molecules is to absorb light energy that the chlorophyll molecules would "miss," and to transfer it to them, and ultimately to the trap.

We can now comment on the energy-collecting efficiency for light of various wavelengths. Light at wavelengths much longer than 6700Å consists of photons with too little energy to excite the chlorophyll molecules, so none of this energy is collected. Light at wavelengths shorter than 6700Å consists of photons with more than enough energy to excite the chlorophyll molecules to the 1.88-eV state. When these wavelengths are absorbed, the excess energy is wasted. Photons from light near 6700Å have just the right energy for exciting chlorophyll molecules. This energy is collected with optimal efficiency. However, even in this case, a small amount of energy (the difference between the chlorophyll molecule excitation energy of 1.88 eV and the trap excitation energy of 1.80 eV) is wasted. Note that the overall photosynthetic efficiency (35 percent) given in Eq. 3.4 referred to this optimum wavelength region.

Oxidation of H_2O to O_2

The oxidation of H_2O and the formation of NADPH proceeds by a long chain of oxidation-reduction reactions, each of which consists of the transfer of a hydrogen atom from one molecule to another. One version of this process, known as the *series scheme,* is shown in Fig. 3.1. The arrows are to be interpreted, "An electron (and a proton) are transferred from the one side to the other side." The chain of reactions must occur twice, i.e., two hydrogen atoms must be transferred, for the oxidation of H_2O to $\frac{1}{2}O_2$ and for the reduction of NADP to NADPH.

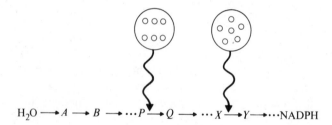

Figure 3.1

As with any chemical reaction, these can proceed in either direction. Again as with any chemical reaction, they tend to go in the direction of lower chemical energy (technically, free energy), i.e., they "run downhill." Since the main purpose of photosynthesis is to achieve a state of higher chemical energy, i.e., to "get uphill," energy must be added to the chain at some point. It turns out that energy is added at two points, between P and Q and between X and Y, as indicated in Fig. 3.1. (The wavy arrows from the dotted balloons are to be interpreted, "Light energy collected by the photosynthetic unit is added here.") It is at these points that the light energy held in the trap enters the chemical reaction.

One model of the chain of reactions is shown on an energy scale in Fig. 3.2. Note that this scale is the free energy, not the total energy. (Since energy is conserved, the total energy clearly does not change when a chemical reaction occurs.) The two upward steps are each about one electron-volt free energy. This

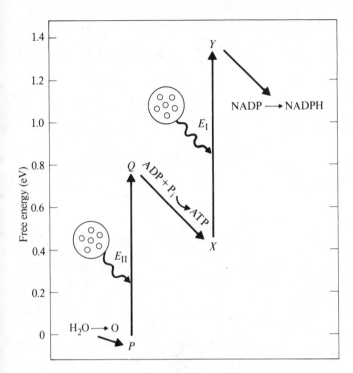

Fig. 3.2 Free energy scale for the reaction chain of Fig. 3.1.

energy is supplied by the trap. One of the two downward sloping regions is used for the conversion of ADP to ATP. The gain in free energy of the ATP is only slightly less than the loss in free energy along the chain, i.e., the process is quite efficient. The function of the other downward-sloping region will be discussed shortly.

Since there are two upward steps in the chain, two photons are required for each passage through the chain. Eight photons will take four hydrogen atoms through the chain, forming two molecules of NADPH and four molecules of ATP,[3] and oxidizing two molecules of H_2O to O_2.

Reduction of CO_2 to CH_2O

The reduction of CO_2 to CH_2O takes place through a series of chemical reactions, collectively known as the Calvin Cycle. The overall stoichiometry is

$$3ATP + 2NADPH + CO_2 \rightarrow CH_2O + H_2O + 3ADP + 3P_i + 2NADP. \qquad (3.12)$$

ATP and NADPH revert back to their low-energy constituents; the energy thus made available fixes one atom of carbon. This cycle takes place in the body of the cell, being supplied with ATP and NADPH from all the photosynthetic units.

The three stages of the process are brought together in Fig. 3.3. (Note that the second stage outputs an excess of ATP.[3] This extra ATP provides an energy

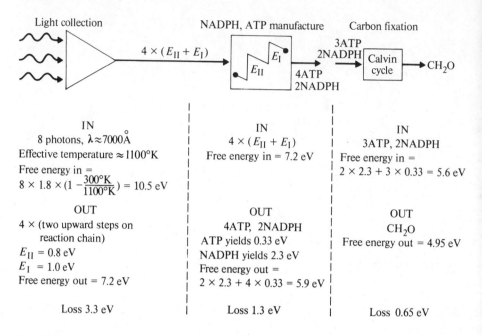

Fig. 3.3 Three steps in photosynthesis. The energies indicated are free energies at standard concentration.

source for cellular processes other than photosynthesis.) In order to treat the stages separately, we have ignored one important consideration—chemical concentration. The free-energy changes indicated in the figure are for standard concentration of all components. In operation, actual concentrations may be quite different, and therefore so will the free-energy changes (see discussion at the end of Appendix A). For example, the light-collection stage can supply more free energy per upward step (E_{II}, E_I) than is required at standard concentration. Those reaction steps can therefore be driven far "to the right," causing increased concentration of products, reduced concentration of reactants, and causing an increase in E_I and E_{II}. Similarly, the "unused" downward slope after step E_I could cause reactions along it to go hard to the right, resulting in high concentrations of NADPH. This high concentration would speed the transport of NADPH from the photosynthetic site to the body of the cell, by diffusion, and cause the Calvin Cycle to run more to the right. Under operating conditions, the loss in free energy of the light-collection stage could thus be *less* than at standard concentration, and the loss in free energy of the Calvin-Cycle stage could be *greater* than at standard concentration.

Considering the overall process, we see that about half of the free energy supplied by light is converted to carbohydrate, and the other half is "wasted." Is that doing well or poorly? We first note that if a chemicel reaction $A \to B$ proceeds with no loss in free energy, then the reaction $B \to A$ will simultaneously proceed at an equal rate. In order to obtain any *net* production of B, there must be some loss in free energy. The overall system will proceed downhill or not at all! We further suspect that if the loss in free energy is very small (the downhill slope very gentle), the net production of B will be very slow. A useful analogy is a battery with internal resistance. If energy is withdrawn from the battery at a very low rate (low current), the power loss internal to the battery is a small fraction of the power delivered to the external circuit, but the power delivered is itself small. As more power is delivered to the external circuit, a larger fraction is lost internally. It is easy to show that when the maximum amount of power is being delivered to the external circuit, an equal amount of power is dissipated internally. This result—that the maximum rate of useful energy extraction occurs when an equal amount of energy is being wasted—is fairly general and applies to any linear system (see Problem 3.4). Photosynthesis uses half and dissipates half of the available free energy, and therefore extracts energy as effectively as an optimized linear system.

3.3   Typical Photosynthetic Efficiencies

The 35-percent efficiency figure mentioned in Section 3.1 is for *gross* production under *ideal* conditions using *red light*. We will now translate this to a figure for *net* production under *actual* conditions using *solar radiation*.

Solar Radiation vs. Red Light

As discussed in Section 3.2, the photosynthetic efficiency is maximum for red light near 6700Å, drops rapidly to zero for lower-energy photons, and falls slowly with higher-energy photons. We can estimate a correction factor for the frequency distibution of solar radiation with the following simple model. Assume all photons with energy $E > E_0$ (frequency $\nu > \nu_0$) are absorbed, and ultimately lead to an excited energy level E_0. Further assume that photons with energy less than E_0 are not absorbed. Then the efficiency $\epsilon(\nu)$ has a very simple dependence on frequency:

$$\epsilon(\nu) = \left(\frac{\nu_0}{\nu}\right)\epsilon(\nu_0) \qquad \nu > \nu_0 \qquad (3.13)$$

$$\epsilon(\nu) = 0 \qquad\qquad\quad \nu < \nu_0$$

If the incident light is distributed according to some frequency distribution $N(\nu)d\nu$, then the average efficiency $\bar{\epsilon}$ is given by the expression

$$\frac{\bar{\epsilon}}{\epsilon(\nu_0)} = \frac{\int\limits_{\nu_0}^{\infty} h\nu_0\, N(\nu)\, d\nu}{\int\limits_{0}^{\infty} h\nu\, N(\nu)\, d\nu}. \qquad (3.14)$$

The denominator is the total incident light energy; the numerator is the usable incident light energy. Of interest to us is solar radiation, adequately described by the Planck black-body distribution (Appendix C), with $T = 5800°K$. Evaluation by numerical integration yields the value $\bar{\epsilon}/\epsilon(\nu_0) = 32$ percent.

As an aside, we can ask the question: Is the location of the chlorophyll molecule's first excited state (1.88 eV) optimum from the point of view of light-collection efficiency? If it were lower, energy could be collected over a broader range (farther into the red), but the higher frequencies (blue light) would be less efficiently used. Conversely, if the energy level were higher than 1.88 eV, blue light would be more efficiently used, but some of the red portion of the spectrum would be lost. In Fig. 3.4, $\bar{\epsilon}/\epsilon(\nu_0)$ is plotted *vs.* assumed molecular excitation energy E_0. The maximum of $\bar{\epsilon}/\epsilon(\nu_0)$ does not occur at the location of the chlorophyll molecule's first excited state. Has nature made a mistake? [1] A little reflection suggests we have optimized the wrong thing—nature would optimize $\bar{\epsilon}$, not $\bar{\epsilon}/\epsilon(\nu_0)$. As ν_0 and E_0 are lowered, the temperature of the isotropic solar radiation field drops (see Eq. 3.9), and $\epsilon(\nu_0)$ falls. A plot of $\bar{\epsilon}$ *vs.* E_0 is also shown in Fig. 3.4. The maximum is reasonably close to the chlorophyll's first excited state, considering the crudeness of our model.

We must correct our simple model because not all light incident on a plant is absorbed. The reflectivity of leaf cover is about 20 percent, further reducing the efficiency.

Finally, we must consider how photosynthetic efficiency depends on light intensity. At low intensity, too few traps are excited, and the chain of chemical reactions does not proceed to the right because of the low concentration of filled traps (see discussion on concentration at end of Section 3.2). As the light intensity increases, the efficiency rises until the light intensity is sufficient to keep most of the traps filled. It is at this intensity that the efficiency is optimum, and to which the red-light value of 35 percent refers. As the light intensity increases further, the rate of photosynthesis remains practically constant, and therefore the efficiency falls. With all the traps filled, additional light intensity serves no useful purpose. Now, "ideal conditions" for growing plants means that they have plenty

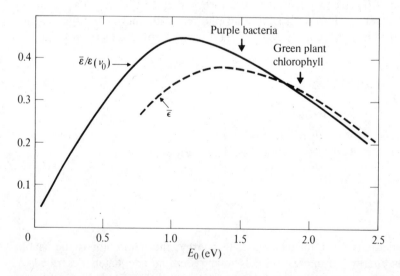

Fig. 3.4 $\bar{\epsilon}/\epsilon(\nu_0)$ and $\bar{\epsilon}$ as a function of assumed molecular excitation energy E_0. The vertical scale for $\bar{\epsilon}$ is arbitrary. The actual excitation energies for chlorophyll from green plants and purple bacteria are indicated.

of sunshine. If the lower leaves receive adequate sunlight during most of the day, then the more exposed leaves will certainly receive more sunlight than they can utilize much of the time. A reasonable "guestimate" is that half the incident sunlight is excess.

Combining the above considerations, we expect the red light, laboratory efficiency of 35 percent to be corrected by a factor of 0.32 for the frequency range of solar radiation, a factor of 0.8 for leaf reflectivity, and a factor of 0.5 for excess light intensity, to predict an efficiency for gross production under ideal field conditions with solar radiation of

$$0.35 \times 0.32 \times 0.8 \times 0.5 = 0.045,$$

or 4.5 percent. The observed efficiency is 4 to 5 percent, in surprising agreement.

Ideal Conditions vs. Actual Conditions

Conditions are usually not ideal. Therefore, typical efficiencies for gross production are less than 4–5 percent, often considerably less. A few examples will illustrate the point. Under desert conditions, because water is in short supply, plants will be thinly spaced, and most of the incident sunlight will strike the ground rather than foliage. Near the surface of the deep ocean, where nutrients (nitrates, phosphates) are in short supply, plant growth is limited. Therefore, the sun's energy is not fully utilized. In the winter, deciduous trees shed their leaves, "hibernate," and use none of the (then limited) solar energy. It is only under ideal conditions of thick vegetation, so that all sunlight strikes leaf surfaces, plenty of water, and plenty of nutrients that the 4-percent gross efficiency is achieved. Some typical gross photosynthetic efficiencies are listed in Table 3.1.

Table 3.1 Some typical gross photosynthetic efficiencies.

Ecosystem	Grams/m²-day dry organic matter, gross
Deserts	<0.5
Grasslands, deep lakes Mountain forests Some agriculture	0.5–3.0
Moist forests Shallow lakes Moist grasslands Most agriculture	3–10
Some estuaries, springs, coral reefs Terrestrial communities on alluvial plains Intensive year-round agriculture (sugar cane)	10–25
Continental shelf waters	0.5–3.0
Deep oceans	<1.0

1 gram dry organic matter $\approx$ 4 kcal
Sunlight = 2000–5000 kcal/m²-day, depending on locale
25 g/m²-day $\approx$ 2.5% efficiency

Source: Adapted from E. P. Odum, *Fundamentals of Ecology* (2nd edition). Philadelphia: W. B. Saunders, 1959, p. 75. By permission.

Gross Production vs. Net Production

A plant will use some of its photosynthetically stored energy for the many things it has to do besides grow. This includes such activities as obtaining water and nutrients, reproducing, and defending itself against insects. The use of energy for purposes other than growth is called *respiration,* whether carried out by plants or animals. (Chemically, it is carried out by the reaction: carbohydrate $+ O_2 \rightarrow CO_2 + H_2O$ + energy for activities.)

Gross production per unit time is the total amount of solar energy converted to chemical energy in that time. *Net production* per unit time is the increase in the amount of stored chemical energy over that time. The difference is respiration: gross production minus respiration yields net production. It is the net production that is available for energy (food) for the next member on the food chain. Net production is always less than gross production, often considerably so.

The net-to-gross ratio varies considerably between ecosystems, being high during early successional stages (such as a burnt-over forest or an abandoned field), and low for climax and near-climax ecosystems (such as an old deciduous forest). Note that the preceding statement refers to the entire ecosystem rather than just the plants in it. Much of man's agriculture is a "pulsed" arrangement, where things start from scratch each year. It is therefore like an early successional stage and has a high net-to-gross ratio.

3.4   Food Chains

Green plants obtain their energy from the sun by photosynthesis (and their carbon from CO_2, also by photosynthesis). Animals obtain their energy (and carbon) by eating plants or by eating other animals that have eaten plants, etc. This chain of eating and being eaten is called the *food chain.*

An idealized food chain is shown as an energy flow diagram in Fig. 3.5. Three "trophic levels" are shown. Level (1) consists of the producers, also known as *autotrophs.* For our purposes, it is synonymous with green plants. Level (2) contains the primary consumers or grazers, the herbivores, or colloquially, the vegetarians. Level (3) contains the secondary consumers or carnivores, or colloquially, the meateaters. Members of all levels other than Level (1) are collectively referred to as *heterotrophs.*

As shown in Fig. 3.5, about one percent of the absorbed solar energy is converted to net plant production; ten percent of the net plant production is converted to net herbivore production; ten percent of the net herbivore production is converted to net carnivore production. These figures are typical and show the substantial losses in energy as one moves up the food chain. There are even longer food chains, with tertiary consumers eating carnivores, etc. However, the amount of energy available at the fourth or fifth level is very small. (The energy transfer at these levels is also about ten percent.) Aquatic food chains tend to have four to five levels, terrestrial ones three to four.

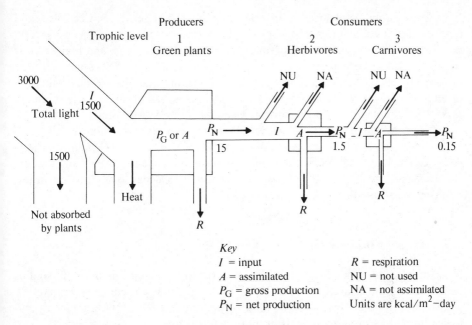

Fig. 3.5 Energy flow in an idealized food chain. (From *Ecology* by Eugene P. Odum. Copyright © 1963 by Holt, Rinehart and Winston, Publishers. Reprinted by permission of Holt, Rinehart and Winston, Publishers.)

In the nonphotosynthetic conversion stages, the 90 percent of energy not transferred is either not eaten, eaten but not assimilated, or assimilated but used in respiration. At first, this ten percent efficiency of energy transfer seems rather low. There are several reasons why it is not higher. (1) Not all of the production at one trophic level is suitable as food for the next level. Some obvious examples are grass roots, tree branches, bird feathers, and animal bones and fur. (2) Some usable food must be left behind to assure a continuing supply. For example, if insects completely strip a tree of its leaves, the tree dies. If more than 50 percent of the grass on a range is eaten, the range deteriorates. (3) Energy is required for activity other than growth, that is, for respiration. Food gathering, protection from predators, reproduction, breathing, blood circulation, cell replacement, etc., all require energy. Energy is required to maintain a biological system in good working order, whether that system is a plant, an animal, or a cell within a plant or animal. (4) As one moves up the food chain, the biological material becomes more complex. Carbohydrate must be converted to protein, for example, and many other specialized molecular parts are required. Energy is required for the chemical conversion. In thermodynamic language, the more complex molecules are more organized, less random. They contain more information, and therefore have less entropy per unit energy, and correspond to a "higher temperature." In their manufacture, the Second Law requires some waste heat, and to the extent that the biological efficiency is below the thermodynamic limit, there will be more waste heat. This waste heat appears as respiration in the energy budget.

Some examples of food chains are grass-antelope-tiger, leaf-insect-bird, and phytoplankton-zooplankton-fish-bigger fish. A given species may occupy several positions in a food chain. For example, bears eat berries, grubs, and honey; birds eat seeds and insects; perch eat water plants, zooplankton, and small fish; man eats vegetables, beef, perch, and tuna fish. The actual quantitative determination of all the energy flows in any specific ecosystem is a difficult and time-consuming task. The complexity of most natural systems is quite bewildering.

The food chain just discussed is called the *grazing food chain,* that is, food is eaten live or freshly killed. In addition, there is a *detrius food chain,* that is, a decay chain. Plant and animal growth not consumed while living dies and decays. In the decay process, food (energy) is furnished to decay bacteria, soil mites, millipedes, and the like.

3.5   Man's Food Chains — Agriculture

Most of modern man's food comes from agriculture. Fishing is the only significant source from a natural food chain. In this section, we compare agriculture with natural food chains to see in what ways agriculture is an improvement, and to see how modern agriculture is an improvement over older practices.

Gross Photosynthetic Productivity

Agricultural ecosystems usually yield significantly *lower* annual gross productivity than natural ecosystems on similar terrain. This is true because (1) leaf cover is much thinner, and (2) leaf cover is present for a smaller fraction of the year. There are occasional exceptions to this generalization, for example, when barren land is rendered productive by irrigation or fertilizer (i.e., when a crucial element had been missing). It should be stressed that any advances in agriculture are not due to more efficient use of sunlight, in the sense of an increase in the efficiency of the photosynthetic process.

Net-to-Gross Production Ratio

This ratio is considerably higher for agriculture, and higher for modern than for older agricultural practices. There are two reasons for this. (1) Most agriculture is "pulsed," that is, it starts fresh each year. In this sense it is like an early successional stage, and possesses the high net-to-gross production ratio characteristic of most early successional stages. (2) The net-to-gross production ratio is increased (i.e., respiration is reduced) by man doing jobs for plants and livestock that in nature they would do for themselves. In natural ecosystems, plants must compete with other plants for water, sunlight, and soil nutrients; they must resist insects, disease, and herbivores in general; they must have a system of seed dispersal; etc. All these jobs require the plant to use energy, i.e., to respire. In agricultural ecosystems, man plows, fertilizes, weeds, uses insecticides, fences out large herbivores, and sows seeds. All these jobs require *man* to use energy, which in modern agriculture invariably comes from fossil fuels. In effect, food is now raised partly on energy from sunlight and partly on energy from fossil fuels. The situation with livestock is similar. Food is supplied, predators are warded off, young are cared for. Work that in natural ecosystems is done by the animal is performed by man in an agricultural ecosystem, again largely with fossil-fuel energy.

Useful Production

The fraction of the net production that is useful to man (namely, edible) is higher for agriculture than for natural ecosystems, and has been increased by modern agricultural techniques. This has been accomplished in two ways. (1) Only the desired species of plants are raised. (Seeds are sown for the desired species; other species are removed by weeding.) (2) The "useless" components of plant and animal species have been eliminated (through selective breeding), with man performing the functions formerly done by the eliminated components. Again, man must use energy to do this, and it invariably comes from fossil fuel. For example, seeds have been eliminated from some varieties of oranges as a "useless" inedible part. The species is propagated by grafting. (Item (2) of this paragraph is quite similar to item (2) discussed in the preceding paragraph.)

In summary, progress in agriculture has been through man's doing part of the plant or animal's job, and through selective breeding of plants and animals that would take advantage of man's efforts. In a natural situation, these plants and animals would fare very poorly. They require considerable help, considerable "fossil-fuel subsidy." For this reason, modern U.S. agricultural practices cannot be readily "exported" to underdeveloped countries. The missing ingredient in the underdeveloped countries is not the lack of information, but rather the lack of sufficient nonsolar energy to apply to agriculture, namely, the fossil-fuel subsidy.

Questions and Problems

3.1 † ‡
In Section 3.1, it is asserted that by focusing unscattered light from a radiant object (e.g., the sun), one can in principle achieve an energy flux as high as that leaving the surface of the radiant object, and can therefore heat something placed in the image plane to a temperature equal to that of the emitting radiant object. Demonstrate this by an example.

Warnings and hints: This is not an easy problem, because the aberrations in the focusing system enter in an essential way. Consider using a paraboloid of revolution to focus the light from a distant radiant disc. If you choose the aperture of the paraboloid large enough so that light approaches the image from 2π steradians, and dare to ignore aberrations, you will find an incident flux *twice* as large as that leaving the radiant object, in apparent violation of the Second Law. If instead you use a small aperture so that the neglect of aberrations is justified, then you will find that the flux in a small solid angle about normal incidence on the image equals that in a similar solid angle leaving the radiant object. This is a correct result, but a limited one. To demonstrate that the flux incident on the image disc over a full 2π solid angle equals the flux leaving one surface of the radiant object disc (the desired complete demonstration), it is necessary to take aberrations into account, which is a difficult task.

The intent of this problem is to convince you that the temperature of an unscattered radiation field is the same as that of the object that emitted the radiation. However, unless you are quite skilled in geometrical optics, you may come to quite a different conclusion!

3.2 * ‡

Give a derivation for the temperature of diffuse solar radiation not using the concept of entropy (as was done in Section 3.1). Instead, make use of a hypothetical material that is perfectly reflecting at all frequencies except for a narrow band $d\nu$ around some frequency ν. In this narrow region, the material is assumed to be perfectly absorbing Compare the result you obtain with Eq. 3.8; explain the small discrepancy.

Hint: Place an object coated with this material in a radiation field emitted by a distant source (the sun) at temperature T_0, and then rendered locally isotropic by scattering. Require that the energy emitted by the object equal the energy absorbed by it. Use the Planck distribution law to express the energy emitted by the source and the energy emitted by the object. Identify the temperature of the object with the temperature of the radiation field at that frequency.

3.3

Pretend you are God designing photosynthesis (or an engineer designing a solar energy system, if you prefer). Weigh the relative merits of a system that takes advantage of the "point source" nature of solar radiation (a focusing system) on the one hand, and a system that works on diffuse solar radiation on the other hand. Make semi-quantitative estimates of the probable relative efficiencies of the two systems, and of the probable on-times of the two systems. (*On-time* is the fraction of the time that the system receives solar energy of a variety that it can utilize.) Discuss other relevant considerations.

3.4

The following are examples of systems with transfer of energy and dissipation of energy. The rate of useful energy transfer is compared with the rate of dissipation.

a) A battery with emf $\mathcal{E}$ and internal resistance r is connected to a variable external load R, as shown in the figure. Define the efficiency of the system as the ratio of the

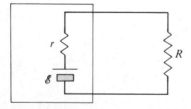

power delivered to the external load to the total power dissipated (internal plus external). Plot on a graph the efficiency, the power delivered to the external load, and the power dissipated internally as a function of R. Show that when the power delivered to the external load is maximum, the efficiency is 50 percent, i.e., the power dissipated in the internal resistance r equals that delivered to the external load R.

b) Gas under pressure P in a cylinder of cross-sectional area A does work when it expands against a piston being withdrawn at a velocity v. Some energy is dissipated

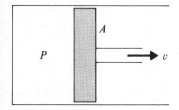

due to the friction between piston and cylinder, and the rest goes into useful work. The frictional force is assumed to be proportional to v. If the velocity is chosen to give the maximum rate of useful work (maximum useful power), show that useful power equals dissipated power.

c) Consider a generalized (constant) force F, a generalized displacement x, and a generalized frictional force $f = \alpha \cdot (dx/dt)$. The rate at which the force F does work is $F \cdot (dx/dt)$. Of this, $f \cdot (dx/dt)$ is dissipated, and $(F - f) \cdot (dx/dt)$ is power for useful work. Prove that if dx/dt is chosen so as to give maximum useful power, then used power equals wasted power, and efficiency is 50 percent.

d) Give additional examples of systems with energy transfer and dissipation. When the appropriate current, flow, velocity, etc., has been chosen so as to maximize the rate of useful energy transfer, does useful power equal dissipated power?

3.5

With modern "chicken factory" agriculture, 1.5 pounds of dry corn (4.5 kcal/g) can be converted to one pound of chicken (5 kcal/g dry weight, 70 percent water). With what efficiency is energy being converted from corn to chicken?

In marked contrast, a human increases body weight from zero to 150 pounds in 15 years on a diet of 2500 kcal/day. With what efficiency is energy being converted from food to human? (Again assume body weight is 70 percent water and 30 percent dry weight, and use 5 kcal/g for dry animal organic matter.)

You'll find your second answer is much smaller than your first one, and also quite a bit lower than the 10-percent figure quoted in Section 3.4 for conversion of net production at one level in a food chain to net production at the next level. This second difference is even more marked, since a large amount of net production at the level from which humans consume is *not* included as ingested food, but ends up as agricultural waste, food processing plant waste, or garbage. Make some conjectures as to why the efficiency for net human production is so low. (The note in Problem 3.7 may be of some help.)

3.6

Good New Zealand grassland will support one 1000-pound cow per acre; the cow in turn produces 4000 pounds of milk per year. Dry vegetable organic matter contains 4 kcal/g; milk is 88 percent water, 4 percent fat (9 kcal/g), and 8 percent nonfat solids (4 kcal/g). With what efficiency is net green plant production being converted to net herbivore production (in the form of milk)? For purposes of this problem, assume the grassland has a gross yield of 5 grams of dry organic matter per square meter per day (see Table 3.1), and that the ratio of net production to gross production for the grassland is 1:2.)

3.7 *

Before its demise in the nineteenth century, the bison was the principal large herbivore on the American Great Plains, the interior grasslands covering about one-fifth of the

U.S. land area. One estimate put the number of bison at 60 million. From energy considerations, does this estimate seem reasonable?

Note: In order to relate standing crop (i.e., total weight) of some species, in this case bison, to net production of that species, it is necessary to know something about the growth rate, breeding pattern, and lifespan of the species. For this problem, assume that a bison reaches a mature weight of 1500 pounds in two years, and that females give birth to one calf each year from their second year on. For a stable population of bison (neither increasing nor decreasing in number), the mean lifespan would then be to 2 to 3 years. (Justify this number.) (This mean lifespan could be composed of each cow reaching 3 years of age and having 2 calves, a male and a female, both of which survive to maturity. It could also be composed of a high infant-mortality rate, with the few cows that survived to the child-rearing age then living much longer than the mean. The relationship between net production and standing crop will depend to some extent on the details of the age distribution, but not strongly.) From the information given, a reasonable estimate for annual net production of bison is 500 pounds live weight per bison. (Justify this number.)

3.8

Evaluate the deep ocean (i.e., the ocean exclusive of continental-shelf waters) as a source of food for humans. Assume the fourth trophic level (second-level carnivores) would be used for human food. (The phytoplankton at the surface of the deep ocean are very small, and the animals that feed on them are also small. Therefore, food chains are long and trophic levels below the fourth are unsuitable for supplying human food.) Assume a 10-percent harvest. (This is optimistic for harvesting a low-density crop spread over such a large area.) How many people could have their caloric needs supplied from this source? For the present world population, what fraction of an individual's caloric needs could come from this source?

Comment: The conclusion that the deep ocean is unimportant for food supply is a correct one. Continental-shelf waters are more favorable. Though the area is down by a factor of ten, net productivity is up by a factor of approximately 4. The plant life is larger and food chains are shorter, so the third trophic level is suitable for supplying human food. Because of the higher density and smaller area, a harvest of 40 percent can be contemplated. (Higher harvesting could not be sustained for many years without destroying the fishery.) Even here, the continental-shelf waters can supply a negligible fraction of human caloric needs. However, because food is taken from high trophic levels, it is rich in protein. Therefore, continental-shelf waters *can* supply a significant fraction of human *protein* needs, perhaps 20 percent of that for the present world population.

Notes

1

Because an escape from teleological thinking has been a major philosophical advance of twentieth-century biology, statements like this will cause the reader with a life-science background considerable discomfort. He can ease the pain somewhat by re-phrasing in terms of the selective advantage one scheme has over another.

2

Some complicated chemicals are better handled by their abbreviations than by their names or chemical formulas. In this section we will encounter the following.

NADP   —   nicotinamide adenine dinucleotide diphosphate
NADPH —   the reduced form of NADP
ATP   —   adenosine triphosphate
ADP   —   adenosine diphosphate
P_i —   inorganic phosphate

These chemicals take part in the following reactions.

$$ATP \rightleftarrows ADP + P_i + 0.33 \text{ eV}$$
$$NADPH \rightleftarrows NADP + 2H + 2.3 \text{ eV}$$

ATP plays an important energy role in most biological systems.

3

The experts do not agree on the amount of ATP produced along with the formation of NADPH. Some say two molecules of ATP are produced per NADPH molecule (as we have indicated in Figs. 3.2 and 3.3), while others say only one molecule of ATP is produced. In the latter case, the second stage would output less ATP than is necessary for the third stage, and the missing ATP would have to be supplied from some other cellular process. However, our conclusions about efficiencies would not be significantly altered.

Related Reading

R. P. Levine
"The Mechanism of Photosynthesis," *Scientific American* **221,** No. 6, p. 58.

Roderick K. Clayton
Molecular Physics in Photosynthesis. New York: Blaisdell Publishing Co., 1965.

H. R. Mahler and E. H. Cordes
Biological Chemistry. New York: Harper & Row, 1966. Chapter 11 covers photosynthesis.

R. S. Knox
"Thermodynamics and the Primary Processes of Photosynthesis," *Biophysical Journal* **9,** No. 11, p. 1351.

Eugene P. Odum
Ecology. New York: Holt, Rinehart, and Winston, 1963. This short book furnishes an excellent quick introduction to ecology.

Howard T. Odum
Environment, Power, and Society. New York: Wiley-Interscience, 1971. An ecologist, the author is the originator of energy-flow studies in natural ecosystems. Here he looks at energy more generally and presents several novel points of view. Section 5 of this chapter leans heavily on Odum's ideas, as expressed in Chapter 4 of his book.

John S. Steinhart and Carol E. Steinhart
"Energy Use in the U.S. Food System," *Science* **184,** No. 4134, p. 307. A quantitative treatment of the "fossil-fuel subsidy" to U.S. agriculture. For each calorie of food delivered in 1970, two calories of nonsolar energy were input on the farm, three in the food processing industry, and three in commercial and home food-related activities.

4 Sources of Energy

There are two spiritual dangers in not owning a farm. One is the danger of supposing that breakfast comes from the grocery, and the other that heat comes from the furnace.

A. Leopold, *A Sand County Almanac**

4.0 Introduction

In Chapter 3 we discussed where breakfast came from; in this chapter we will consider where heat comes from, that is, we will consider sources of nonbiological energy. There is no single "best" source of energy. Indeed there is no single dominant criterion for determining "goodness." Rather there are several figures of merit whose relevance varies from situation to situation. Even after deciding upon which figures of merit to emphasize, several difficulties remain in evaluating an energy source. Before considering specific energy sources, we will first discuss some figures of merit and itemize some of the difficulties to be encountered.

4.1 Figures of Merit for Energy Sources

Two intrinsic characteristics of an energy source, *concentration* and *quality,* play important roles in determining "goodness." Concentration can be defined as energy per unit volume or energy per unit mass. For a flowing energy source such as solar energy, the relevant measure is energy flux, i.e., energy crossing a unit area per unit time. Other factors being equal, the higher the concentration,

---

* From *A Sand County Almanac, With Other Essays on Conservation from Round River* by Aldo Leopold. New York: Oxford University Press, 1966, p. 6. By permission.

the better the energy source. Extraction, transportation, storage, and handling are easier and cheaper, and a high energy conversion rate per unit volume (needed for many industrial processes) is more readily achieved.

Quality is not so easily defined. As used here, the term refers to the fraction of the total energy that is available for use. The energy per unit entropy or "characteristic temperature" measures this fraction. Consider a unit of fuel with internal energy U and entropy S. If we use that fuel to obtain work W, we know that W will be less than U. Since the entropy of the mechanical work W is zero, and since the entropy of a closed system cannot decrease, the entropy of the fuel must go somewhere, say to a heat reservoir at some temperature T. The energy wasted will be at least TS, the minimum amount of heat flowing to the reservoir. If we define the characteristic temperature of the fuel as $T_c = U/S$, then

$$\frac{\text{Waste energy}}{\text{Total energy}} \geqslant \frac{TS}{U} = \frac{T}{T_c}.$$

Comparing this result with the Carnot cycle (Appendix A), we see that fuel with a characteristic temperature T_c can (in principle) provide useful work as efficiently as a heat reservoir having temperature T_c.

It is useful to consider the microscopic basis for quality. The basis is a high energy per system, where "system" means the smallest entity possessing a complete, organized unit of energy (e.g., energy per bond for chemical energy; kinetic energy per molecule for thermal energy; energy per photon for radiant energy). If E is the energy per system, then one can define a characteristic temperature $\theta_c = E/k$, where k is Boltzmann's constant. For many cases, $\theta_c \approx T_c$.

If energy is extracted from fuel by burning, the actual combustion temperature cannot exceed T_c. If it could, it could be used to drive a Carnot engine at that temperature, thus providing work more efficiently than the upper limit we have obtained. As the combustion temperature approaches T_c, the combustion reaction no longer "goes to completion," i.e., the inverse reaction begins to occur at a significant rate. As a practical matter, fuels are invariably burned well below their characteristic temperatures, with the result that they provide less than the maximum possible amount of useful work. As long as the characteristic temperature is much higher than "room temperature" ($\sim 300°K$), it usually matters little what the actual characteristic temperature is. For example, the very high characteristic temperature of nuclear fuels is *not* taken advantage of; nuclear fuels are used slightly less efficiently than fossil fuels, which have a considerably lower characteristic temperature. The low characteristic temperatures of geothermal energy *is* a disadvantage, as is that of unfocused solar energy.

Measures of the concentration and quality of several energy sources are listed in Table 4.1. (Note the wide range.)

In addition to the two intrinsic energy source characteristics, there are several practical measures of "goodness," the most all-inclusive and readily quantifiable one being cost. Other considerations include total available supply, ease of extraction, ease of transport, and ease of use. Clearly, all these factors are reflected to some extent in cost, and they are determined in part by the intrinsic characteristics.

Table 4.1   Measures of concentration and quality of various energy sources.

Energy source	Energy/volume (kWh/m³)	Energy/mass (kWh/kg)	Energy flux (kW/m²)	Characteristic temperature (°K)
Coal	10,000	7.6	—	} ~10^4
Oil	10,700	12.0	—	
Natural gas	11.0	15.0	—	
Uranium-235	4.4×10^{11}	2.3×10^7	—	~10^{11}
Deuterium	1.2×10^7	6.7×10^7	—	~10^{10}
Geothermal: earth's average	—	—	6×10^{-5}	300
Geothermal: point source	—	—	~10^5	500
Sunlight	~10^{-12}	—	~1.0	1000–6000
Hydropower (30-meter head)	0.8	8×10^{-4}	—	∞

4.2   Problems in Evaluating Energy Resources

With many energy resources, a serious problem is encountered in determining what is there. This applies in varying degrees to all underground sources—coal, oil, natural gas, uranium, and geothermal energy.

The usefulness of a resource depends on the state of technology for extracting, transporting, and utilizing that resource, and also on the economic conditions affecting that resource and other competing resources. Technological and economic changes can make a resource more or less useful. A resource is usually evaluated "under present economic and technological conditions." This means that only the cheapest resources are counted. When these are exhausted, the price rises, and other resources become "available."

Some energy resources have alternate, nonenergy uses. This applies particularly to hydroelectric power resources, for which there are often competing recreational and aesthetic uses. (Water flowing over Niagara Falls—an aesthetic use—is not available to generate electricity.) In the same vein, if environmental damage in extracting a resource is severe (as in the case of strip mining, for example), then alternate uses (those uses eliminated by the environmental damage) compete with energy production. The usual procedure in evaluating an energy resource is to ignore all the competing uses, thus overestimating the available energy resource.

Environmental problems in using an energy resource must be considered. For example, some coal has a high sulfur content which causes severe air pollution problems. Such coal therefore cannot be considered an energy resource for many metropolitan areas.

4.3   Continuous Sources of Energy

Energy sources that are replenished about as fast as they are used are classified as *continuous sources*. Those that are replenished much slower, or not at all, are classified as *stored sources*. Direct solar energy, water power, and tidal energy are examples of continuous sources. Uranium and fossil fuels are examples of stored sources. Uranium is not replenished. Fossil fuels are perhaps being replenished, but at a rate very much slower than they are being used. (Present fossil fuels were deposited over a time span of 6×10^8 years; they are being removed over a time span on the order of 10^3 years.) Fuel wood represents a source that may fall into either classification. If wood is harvested at about the same rate that new growth occurs ("sustained yield," in lumbering terminology), then fuel wood is considered a continuous source. However, if wood is harvested significantly faster than new growth occurs ("mining" lumber), then fuel wood is viewed as a stored source, one that is being depleted. Other factors being equal, a continuous source is preferable to a stored source, because the continuous source entails no thermal pollution (see Chapter 6, Part B), and because exhaustion of supply need not be of concern.

Figure 4.1 shows the earth's energy flows. There are three original sources: (1) tidal energy, (2) geothermal energy, and (3) solar radiation. For comparison, present worldwide human energy use is at a rate of 6×10^{12} watts.

Tidal Energy

The source of *tidal energy* is the combined kinetic and gravitational potential energy of the earth-moon-sun system. By studying changes in the earth's period of rotation, the rate of dissipation of tidal energy has been determined to be

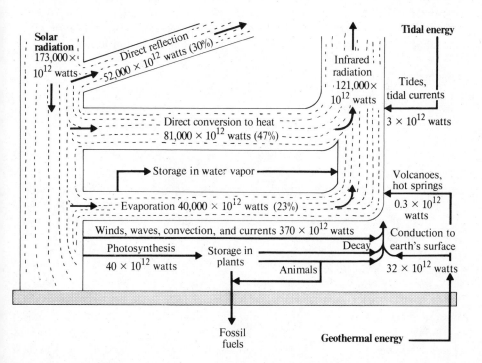

Fig. 4.1 Earth's natural energy flows. (From M. K. Hubbert, "U.S. Energy Resources, A Review as of 1972.")

about 3×10^{12} watts. Of this, about one-third is dissipated in shallow seas. The rest is dissipated in the deep oceans or within the earth's interior. Only that portion dissipated in shallow seas is in any sense available to be harnessed. More realistically, only sites where there is a natural coastal basin and a large rise and fall of tides are suitable for harnessing tidal energy with presently conceived technology. (In this case, a dam would be built between the basin and the open sea, allowing the flow into and out of the basin to be controlled. Flow would be allowed only at low and high tide. Specifically, at low tide with water levels of basin and open sea equal, the flow gates would be closed. The level of the sea would rise with the tide, but the level in the basin would not. At high tide, the gates would be opened, water would flow into the basin, driving generators, and bring the level in the basin up to that of high tide. The gates would then be shut. The level of the sea would fall with the tide, but the level in the basin would not. At low tide, the gates would be opened, water would flow out of the basin, again driving generators, and bring the level in the basin down to that of low tide. The gates would then be closed, and the cycle repeated.) A survey of such sites showed 6.4×10^{10} watts of power available. Even if this power could be captured with efficiencies comparable to present energy use efficiencies (an optimistic assumption), only one percent of current energy demands could be supplied. Tidal energy is therefore too small an energy resource to be important except on a local level. At present, only one tidal electric plant of any appreciable size has been built. It is a 240-MW plant on the Rance estuary in France, and began operating in 1966. There has been extended consideration (since the 1930's) of a project utilizing Passamaquoddy Bay, on the United States-Canadian border off the Bay of Fundy, where the tidal range averages 5.5 meters. As of 1974, spurred by the "energy crisis," the project was again being considered.

Geothermal Energy

Geothermal energy is generated in the earth's interior, principally as a result of decay of radioactive nuclei. The heat flows out from the interior to the surface by conduction, with an average flux of 0.063 watts/m^2 and a relatively low temperature. (For comparison, the flux of solar energy is 1400 watts/m^2, and the characteristic temperature 1300°–6000°K.) This average flow of geothermal energy is too dilute and of too low quality to be of any significance. A small fraction of the geothermal energy (about 1 percent, or 0.3×10^{12} watts) comes to the surface through convection via volcanoes, hot springs, and geysers. These point sources supply energy in a much higher concentration, though still of low quality. The total magnitude of the resource (0.3×10^{12} watts) assures that it will not play a major role as an energy source.

Before giving up on geothermal energy, two related points should be considered. (1) Geothermal energy can be "mined" rather than "harvested," i.e., it can be extracted at rates considerably faster than the natural rate for "short" periods of perhaps 50 years. (2) Small-scale, man-made, geological modifications (perhaps using nuclear explosives) may create new point sources of geothermal energy if the initial geological conditions are favorable. Proceeding along this line, geothermal energy becomes a stored source rather than a continuous one. There is wide divergence of opinion as to how large the resource is, that is, how much energy can be usefully extracted. The uncertainty stems largely from lack of geological knowledge, but also from uncertainties about the success of "geological engineering." Estimates for U.S. electrical power production from geothermal sources in the year 2000 vary from 4×10^9 watts to 4×10^{11} watts. (Current U.S. total electrical power production is 2×10^{11} watts.)

Aesthetic uses compete with energy extraction for geothermal fields with obvious surface indications such as geysers. The geothermal field in Yellowstone National Park, for example, has been reserved for aesthetic purposes.

There are geothermal electric generating plants presently operating in Italy (Larderello), New Zealand (Wairakei), northern California (The Geysers), and a few other locations. Total capacity in 1972 was 900 MW. In most cases, steam is extracted from wells drilled into the steam-bearing rock (see Fig. 4.2), and used to drive fairly conventional generators (see Fig. 4.3). It is estimated that energy is being withdrawn at these sites from 10 to 100 times faster than the natural replenishment rate, so they are not continuous sources. However, their useful life is estimated to be at least 30 to 50 years. The existing plants are quite successful, demonstrating the practical usefulness of geothermal energy. These sites are, however, the most favorable ones.

In summary, geothermal energy is not significant (on a worldwide scale) as a continuous source of energy, and it is simply not known whether or not it can be significant as a source of stored energy. A large research program is being called for to find out.

Solar Energy

The magnitude of the *solar energy* resource is huge. Incident upon the earth are $173,000 \times 10^{12}$ watts, which is 30,000 times man's current power use. Solar energy, therefore, merits very careful consideration. One can consider capturing solar energy directly or indirectly after it has undergone one or more natural transformations.

Direct solar energy
Consider three figures of merit of direct solar energy. Its concentration is very low—the solar constant is 1.4 kW/m². Its quality is adequate—as discussed in detail in Section 3.1, its characteristic temperature ranges from 1100°K to 5800°K. The adequacy of supply is very good—a scheme utilizing one percent of the land surface and capturing energy at one percent efficiency would satisfy man's current power uses. With the low concentration, the problem is "how to do it at a reasonable price." This will be considered in Section 5.4.

Indirect solar energy
Indirect sources of solar energy include winds, waves, and ocean currents, water power, and photosynthesis. Wind has been used as an energy source for several centuries. The total amount of solar power going into winds, ocean currents, etc.,

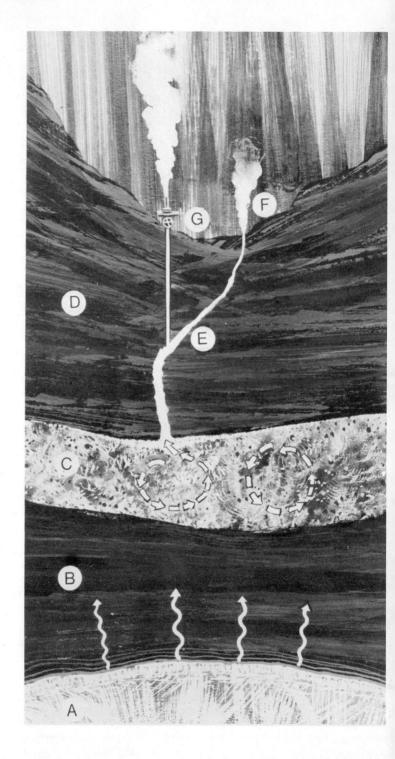

Fig. 4.3   Geothermal steam-electric power plant at The Geysers, California. Piping brings steam from wells to the power plant. Loops in piping allow for thermal expansion and contraction. The building houses two generators with a combined capacity of 54 MW. Cooling is accomplished by forced-draft evaporative cooling towers. As of 1973, the total installed capacity at The Geysers was 400 MW. (Photo courtesy of the Pacific Gas and Electric Company.)

is substantial: 370×10^{12} watts, about 60 times current human power use. However, these sources are (with occasional local exceptions) very diffuse, much more so than direct solar energy. It is hard to conceive of a scheme that could capture anywhere near one percent of these sources. Therefore, they are not a significant worldwide energy resource.

Water power has also been used as an energy source for several centuries. A typical large hydroelectric installation is shown in Fig. 4.4. The advantage of

◀ **Fig. 4.2**   Schematic view of a geothermal field (not to scale) showing (A) magma (molten interior of earth), (B) impervious rock, with heat transported by conduction, (C) porous rock, with heat transported by movement of water, (D) impervious rock, (E) fissure in impervious rock, allowing steam to escape, (F) geyser, fumarole, or hot spring, and (G) well, tapping steam in fissure. (Photo courtesy of the Pacific Gas and Electric Company.)

Fig. 4.4 Noxon Rapids hydroelectric installation on the Clark Fork River in Montana. Four 100-MW turbines and generators (cylindrical objects, right center) have been installed and there are provisions for a fifth. The dam is 190 feet high and 5850 feet long. (Only the central portion is shown—earth fill dams extend to the left and right of the powerhouse and spillway sections.) The dam backs water 38 miles upstream (to the tailwater of another dam) and creates a 8650-acre reservoir. (Photo courtesy of the Washington Water Power Company and the Edison Electric Institute.)

water power is that natural processes (the hydrological cycle) increase the concentration of energy considerably. Needless to say, a price is paid for this concentration. The power input to the hydrological cycle is $40,000 \times 10^{12}$ watts, but the world's estimated water power potential is only 2.8×10^{12} watts. This potential is not only highly concentrated, but of a high quality—it can be converted to electricity with small losses. Therefore, one could view all of the 2.8×10^{12} watts as being available. At present, 10 percent of it is being utilized. In the United States, the water power potential is 0.16×10^{12} watts, of which about one-third is currently being utilized. On the debit side, three points must be noted. (1) Water power potential is not evenly distributed and does not always occur where the need is greatest (see Table 4.2). Western Europe and North America currently use energy at a rate considerably in excess of their water-power potential. (2) To develop or exploit water-power potential, it is usually necessary to construct a dam, creating a storage reservoir and an abrupt vertical drop. These reservoirs fill with sediment deposited by the entering streams, so that the useful life of a storage reservoir is on the order of a century. While the water-power

Table 4.2   World water-power capacity, 1970.

Region	Potential (10^3 MW)	Percent of total	Developed as of 1970 (10^3 MW)	Percent developed
North America	313	11	88	28
South America	577	20	13	2
Western Europe	158	6	93	59
Africa	780	27	6	1
Middle East	21	1	2	8
Southeast Asia	455	16	8	2
Far East	42	1	21	50
Australasia	45	2	7	16
USSR, China, and satellites	466	16	40 [a]	9 [a]
Total	2857	100	278	10

[a] Excludes China.

Sources: Potential from M. K. Hubbert, *Resources and Man*, Committee on Resources and Man, National Academy of Sciences—National Research Council. San Francisco: W. H. Freeman, 1969, p. 209. By permission. Development from *United Nations Statistical Yearbook*, 1972, pp. 157–365.

resource is continuous, the mechanism for exploiting it has a finite lifetime. Those exceptions to the need for building a dam and creating a storage reservoir (e.g., the installation by Niagara Falls, which utilizes Lake Erie for storage) are few in number, and are also invariably cases where there is a very strong competing aesthetic use for the resource. (3) As mentioned in Section 4.2, there are competing nonenergy uses for many water-power resources, or alternately stated, there are environmental damages caused by developing a water-power resource. These will be considered more completely in Chapter 6, Part E. At this point, we must comment that the United States and Western Europe, with about half of their water-power resources developed, and with large and affluent populations (and therefore large needs for aesthetics and recreation), should give serious consideration to retaining the remaining half for nonenergy uses.

In summary, water power is currently a significant but not a major source of energy. If energy use grows in the future as projected, water power will account for a progressively smaller fraction of total use. It will continue to be important in favored geographical regions.

Photosynthesis has been used by man as a nonfood energy source for several tens of millennia (fuel wood), and indeed was his first energy source. (Fuel wood supplied over half of the U.S. annual energy needs until 1885.) As mentioned at the beginning of this section, photosynthetically fixed energy constitutes a continuous or stored source, depending on rate of utilization. Fossil fuels are an example of stored photosynthetically fixed energy. Fuel wood has at different periods in history been both a stored and a continuous source. Here we consider using photosynthetic energy as a continuous source, and can therefore draw upon it only at the rate that it is fixed. The rate at which photosynthetic energy is fixed is estimated to be between 30 and 200 × 10^{12} watts, with a best guess for land-based photosynthesis of 40×10^{12} watts. Fifteen percent of this would supply man's present power uses. In addition to fuel wood, agricultural wastes, municipal sewage, and specially raised fuel crops have been proposed as sources of energy. It is highly unlikely that it will be possible to harvest for energy use anything close to 15 percent of the total land-based, photosynthetically fixed energy. There are substantial competing commercial uses (lumber, pulpwood) and also "in place" uses (e.g., wildlife habitat, flood control, wilderness preservation) that argue against a high-percentage harvest. Therefore, photosynthetically stored energy cannot be a significant, continuous source of energy on a global scale. (It remains, of course, an important source in those regions using a small amount of energy per unit land area, i.e., in some of the underdeveloped countries.)

4.4   Sources of Stored Energy

The sources of stored energy are the fossil fuels (coal, oil, natural gas and natural gas liquids, oil shale, and tar sand) and the nuclear fuels (uranium, thorium, deuterium, and lithium). At present, the technology to exploit some of these resources does not exist. The chemical composition of fossil fuels is that of organic matter—carbon, hydrogen, oxygen, nitrogen, and sulfur—but with the proportions considerably modified. In particular, most of the oxygen is gone.

Coal

Coal was the first fossil fuel to be utilized by man. It replaced wood as the major source of energy in the United States during the second half of the nineteenth century, and was the dominant energy source in Europe somewhat earlier. The geology of coal deposits is such that estimates of coal resources can be made with considerable confidence. There appears to be general agreement that Averitt's studies are approximately correct. He estimates that the United States has 1.5×10^{12} metric tons of mineable coal; the USSR has 4.3×10^{12} tons; the world as a whole has 7.6×10^{12} tons (see Table 4.3). To put these numbers in perspective, 1.5×10^{12} metric tons (U.S. mineable reserves) can supply 2×10^{12} watts of power (current total U.S. power requirement) for 800 years. The rest of the world, which as a whole uses twice as much power as the U.S., has four times as much coal. Therefore, the rest of the world could supply twice its present power requirements for 800 years from its coal reserves. (The preceding arithmetic is useful for setting the scale of our coal resources. It does *not* mean that coal will supply our energy needs for 800 years, because (1) U.S. and world power requirements are rapidly growing, at a rate which will double the present requirements in 20 to 30 years, (2) the geographical distribution of coal is uneven, and (3) there are severe environmental problems associated with using coal.)

One of the serious environmental problems associated with using coal is the release of sulfur dioxide (SO_2) into the atmosphere when coal containing sulfur is burned. For this reason, it is advantageous to use coal with a low sulfur con-

Table 4.3   Coal resources of the world.[a]

	10^9 metric tons	Percent of world resources
USSR	4310	56.0
Asia, excluding USSR	681	9.0
United States	1486	19.5
North America, excluding United States	601	8.0
Western Europe	377	5.0
Africa	109	1.4
Oceania, including Australia	59	0.8
South and Central America	14	0.2
Totals	7640	100.0

[a] The table lists "mineable coal," which is taken as 50 percent coal in place.

Source: Paul Averitt, USGS Bulletin 1275, as quoted by M. K. Hubbert (1974), "U.S. Energy Resources, A Review as of 1972."

tent. Two-thirds of all U.S. coal reserves is estimated to be *low-sulfur coal* (coal with less than one percent sulfur content). In particular, the western coal fields are predominantly low-sulfur coal. However, in the bituminous coal fields of Appalachia, where most coal in the United States is currently being mined, low-sulfur coal is limited, and therefore more expensive.

Oil

Oil rose to prominence as an energy source during the first half of the twentieth century. Soon after the start of the industry, it became "common knowledge" that "we are running out of oil." In 1920, the U.S. annual production was 0.5 billion barrels, and the cumulative production to that point was 5 billion barrels. The chief geologist of the U.S. Geological Survey then estimated that remaining recoverable reserves were in the neighborhood of 7 billion barrels, and warned that U.S. oil production would soon decline. By 1960, U.S. annual production had reached 3 billion barrels; cumulative production to that time was 80 billion barrels, and proved reserves had risen to 32 billion barrels. Throughout the intervening period, as production and proven reserves rose, there were repeated warnings that we were running out of oil, but these warnings proved to be premature. Why has this been so? Several partial explanations suggest themselves. First and foremost, oil is difficult to locate. A comparison with coal is helpful. Most coal deposits lie at depths of less than 1000 feet, are continuous over wide areas, and frequently have surface outcroppings. The average oil well is 4500 feet deep, and some are over 20,000 feet deep. There are no surface indications of the presence of oil. As knowledge of the geology of petroleum improved, more oil was discovered. Second, there has been some confusion between proved reserves and resources. *Proved reserves* are deposits known to exist and to be producible under current technological and economic conditions; *resources* include deposits not yet discovered. Proved reserves are the "shelf inventory" of the petroleum industry. As in any industry, it is bad business practice to maintain too large an inventory. The oil industry has typically maintained proved reserves of 10 to 15 times annual production. Because of the high costs involved in exploring for oil, there was little motivation to seek additional reserves beyond that amount. As demand increased, annual production increased, more reserves were required, and the pace of exploration was adjusted upward to maintain the 10-to-15 year "inventory." Confusion between proved reserves and resources led to the popular misconception that there was only a 10-to-15 year supply of oil left. Third, in extracting oil from a field, not all the oil can be recovered. Oil fields consist of porous formations of sand or rock soaked in oil. A substantial fraction of the oil remains stuck in the crevices in the rock, in spite of various efforts to bring it to the surface. The average recovery rate for U.S. wells has been increasing as improved techniques are applied. Current recovery rates are about one-third, and this figure may be expected to increase further with improved technology and increased expenditures. As the recovery rate rises, oil reserves will rise proportionately even without more discoveries. And fourth, oil is being looked for in more places now than previously. Exploration in Alaska and along the continental shelf has broadened the area searched, and increased well depths have added new territory.

These four points largely explain why earlier estimates of U.S. oil reserves were low, and make us wary of accepting present estimates. The point remains, however, that oil is a finite resource, and at *some* time it will become exhausted. Several recent estimates of U.S. crude oil resources have been made, and they cover a broad range, as can be seen in Table 4.4. It is interesting to note that the high estimate (USGS) and the low estimate (Hubbert) involved a similar technique, and we can therefore identify the reason for the discrepancy. In both studies, the oil discovered per foot of exploratory drilling was plotted against the cumulative footage of exploratory drilling, as shown in Fig. 4.5. The area under the curve gives the cumulative amount of oil discovered. To obtain any predictions about future discoveries, it is necessary to extrapolate the curve to the right. In particular, to estimate the ultimate amount of oil to be discovered, one needs the ultimate

Table 4.4   Recent estimates of U.S. crude oil resources (in billions of barrels).

	Cumulative production to 1970	Proved reserves	Potential reserves	Ultimate production
National Petroleum Council	95	40	125	260
U.S. Geological Survey			365	500
M. K. Hubbert			65	200
L. G. Weeks			235	370
C. L. Moore			215	350
Elliot and Linden			315	450

Source: P. K. Theobald, S. P. Schweinfurth, and D. C. Duncan, "Energy Resources of the United States," USGS Circular 650. Washington, D.C.: U.S. Geological Survey, 1972. Modified to exclude natural gas liquids and to refer to the year 1970.

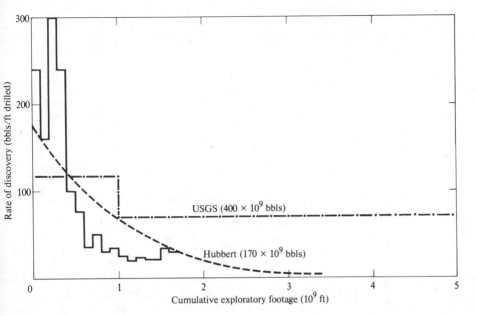

Fig. 4.5 Crude oil discoveries per foot of exploratory drilling *vs.* cumulative exploratory footage in the United States exclusive of Alaska. The dashed curve is the extrapolation used by Hubbert. The dot-dash curve gives the ultimate production estimated by the USGS. (From M. K. Hubbert, "U.S. Energy Resources, A Review as of 1972." USGS curve added by the author.)

amount of exploratory footage. The USGS estimate for this is 5 billion feet, sufficient to explore all favorable strata to a density of one well per 2 square miles. Hubbert also used this figure. The discrepancy comes in their extrapolations of oil discovered per foot of exploratory drilling. The USGS study apparently assumes that future discoveries will be made at rates not far below the average of previous discoveries. But Hubbert notes that the discovery rate has fallen substantially, and assumes it will continue to fall. Since the regions considered to be the most favorable would be explored first, one would expect a decrease in discovery rate. This trend would be counteracted by improvements in technology and geological knowledge, and by exploration in new areas that, though geologically very promising, are difficult to work (e.g., offshore areas). The historical record shows a falling discovery rate, implying the former factor has outweighed the latter ones. Barring a qualitative improvement or "breakthrough" in oil geology or technology, the Hubbert assumption appears more reasonable. Hubbert's estimate is that, of the oil ultimately to be produced within the continental United States, 83 percent has already been discovered. With such a small amount remaining to be discovered, his results are quite sensitive to the recovery rate of oil in fields already discovered. In particular, if secondary recovery techniques were to give substantial yields from fields thought to be exhausted, these yields would be in addition to Hubbert's estimate.

As a working number, we will take 200 billion barrels as an estimate of the remaining oil to be produced in the United States (including Alaska and offshore regions). This number is the geometric mean of Hubbert's estimate (105×10^9 bbl) and the USGS estimate (405×10^9 bbl). If this number is off by a factor of two either way, there is at least one expert who will not be surprised!

Estimates of world oil resources are more uncertain than estimates of U.S. oil resources, and are generally made by geological analogy with the United States. Several studies suggest that recoverable world oil resources are 6 to 10 times larger than U.S. resources, including past production. We will take 2500 billion barrels (or 8.3×300 billion barrels) as our working figure for world oil reserves, bearing in mind that the true figure could be a factor of 2 smaller or more than a factor of 2 larger.

To visualize what these quantities of oil mean, we can perform .the same arithmetic that we did for coal. Two-hundred billion barrels of oil (estimated remaining U.S. recoverable resource) can supply the current total U.S. power requirement of 2×10^{12} watts for 21 years. The rest of the world, with ten times as much oil as the United States, could meet twice its present power requirement for 53 years. Realizing that the United States currently obtains about 40 percent of its energy from oil, and that the energy demand is likely to double in less than 30 years, these figures suggest once again that the United States is running out of oil. Only time will tell whether this is just another false alarm, or whether at last our oil reserves have been depleted.

Natural Gas and Natural Gas Liquids

Natural gas has a geological history of formation very similar to oil, and it is therefore found in similar places. As with oil, it is extracted by drilling a well into a reservoir. Frequently a reservoir will contain both oil and natural gas. As extracted, natural gas is predominantly methane (CH_4), but diminishing amounts of the heavier members of the C_nH_{2n+2} hydrocarbon series—ethane, propane, butane, pentane, etc.—are also present. Those members with $n \geqslant 5$ (pentane and heavier), though extracted from the ground in gaseous form, are liquids at standard temperature and pressure (14.7 psi at 0°C). Propane and butane, which are gases at stp, can be liquefied under modest pressure. It is customary to process "wet" natural gas so as to remove a substantial fraction of the heavier components (propane and heavier). This removed portion is referred to as *natural gas liquids* (abbreviated NGL), and the remaining portion is called *dry natural gas*. NGL contains a component that is liquid at stp as well as a component requiring pressure to liquefy. The latter, referred to as *liquid petroleum gas* (LPG) is commonly encountered as "bottled gas." (While on the subject of abbreviations, dry natural gas is now being liquefied for transoceanic shipment by cooling to low temperatures. This *liquefied natural gas* is abbreviated LNG.)

Since natural gas and oil are found in the same search process and are often extracted from the same field, their development as energy sources has progressed pretty much in parallel. Initially, problems in transporting natural gas limited the use of the resource to regions near the producing fields, and caused some gas extracted from oil wells to be wasted. Long-distance gas pipelines have solved the transportation problems in the United States, and pipelines along with cryogenic tankers for shipment of LNG are doing the same on the international level.

Various estimates have been made of the extent of natural gas reserves. For our purposes, it is sufficient to rely on the similar geology of oil and natural gas. We will use our working figures for oil reserves along with the historic ratio of natural gas discoveries to crude oil discoveries to obtain an estimate of natural gas reserves. Using 6400 ft^3 natural gas per barrel (bbl) oil and 0.2 bbl NGL/bbl oil as the ultimate production ratios, one notes that the ratio of the *energy content* of gas and NGL to that of oil is 1.3 : 1. That is, if for every barrel of oil produced, 6400 ft^3 of natural gas and 0.2 bbl of NGL are also produced, then the latter two will contribute 1.3 times as much energy as oil to the world energy supply. The situation for the United States is complicated by the fact that the resource has already been exploited to a moderate extent, and in the early days much gas was wasted. Hubbert estimates that remaining producible gas reserves are 810 trillion ft^3, compared with his estimate of 100 billion bbls of remaining producible oil reserves. The USGS estimates are 2350 trillion ft^3 of gas and 400 billion bbls of oil. These gas-to-oil ratios are sufficiently close to 6400 cu ft/bbl that our conclusion remains unaltered—natural gas and NGL will contribute essentially the same amount of energy as oil to future energy needs. Therefore, the numerical conclusions in the last paragraph of the oil subsection apply here also.

It is important to stress that estimates of oil and gas reserves, no matter how performed, are not independent. Any lack of geological knowledge that causes an error in one estimate is apt to cause an error in the other. In other words, if the oil estimates are too low or too high, then the natural gas estimates are probably also too low or too high, respectively. When we run out of oil, it is likely that we will also run out of natural gas.

Tar Sands

These are sands impregnated with a heavy crude oil too viscous to allow recovery by flow into wells. Large deposits are known to exist in the province of Alberta, Canada; reserves there are estimated at 300 billion bbls. The largest deposit, along the Athabasca River in northeastern Alberta, contains 88 percent of the total estimated reserves and covers an area of 9000 sq mi. The deposit is at places under 2000 ft of overburden and at other places has surface outcroppings.

Quite recently (1967), the first large-scale mining and extraction plant went into operation in Alberta. Sand is mined, oil removed from the sand (by washing in hot water), and the sand returned to the mine. Since on the average only 15 percent of the sand by weights is oil, nearly ten tons of sand must be processed for every ton of oil obtained. This compares rather unfavorably with coal mining in terms of material processed per unit energy extracted. While other smaller deposits exist in the United States, there is as yet no reliable inventory of U.S. or world tar sands.

Oil Shale

These are carbonaceous shales containing the solid hydrocarbon kerogen. The hydrocarbon content varies from a maximum of 65 gallons of oil equivalent per ton of shale on down. Estimates of world shale oil resources are shown in Table 4.5. While the total resource is very large, only 190 billion bbls were considered recoverable under the economic and technological conditions prevailing in 1965. The most promising U.S. oil-shale deposits are in Wyoming, Utah, and Colorado, in the Green River formation. Here, 80 billion bbls are listed as recoverable under 1965 conditions. So far there has been experimental work with oil shales, but no commercial recovery operation. Four sites for commercial operations were leased by the federal government to private industry in 1974.

To extract oil and gas from oil shale, the shale must be mined, crushed, and heated to high temperatures. The kerogen then transforms into oil, gas, and a solid residue. Finally, the crushed shale residue must be disposed of. As with tar sands, the material processed is typically 90 percent waste, and so the weight of material processed per unit energy extracted is unfavorably large. Various schemes of in-place extraction have been proposed with oil shales, as they have with tar sands. Such schemes would bypass the material handling and disposal problems. So far they have not been shown to be successful. A successful in-place extraction scheme would greatly increase the attractiveness of both oil shale and tar sands as energy sources.

Fission Fuels—Uranium and Thorium

The physics and technology of fission energy is discussed in Chapter 5. There it is pointed out that present-day nuclear reactors—*burners*—use a rare isotope of uranium (^{235}U) as their fuel, while *breeder reactors,* currently in a developmental state, will use the common isotope of uranium (^{238}U) or the common isotope of thorium (^{232}Th) as their fuel. Since the composition of natural uranium is 0.7 percent ^{235}U and 99.3 percent ^{238}U, it makes a considerable difference—a factor of 140—whether uranium fuel is used in a burner or a breeder reactor.

U.S. uranium resources are shown in Table 4.6. As the price per pound increases, lower grade and more inaccessible ores become economically available. The dependence of resource on price is fairly steep. Also shown is the increase in cost of electricity as the uranium cost rises above its present price of $8/lb. Compared with the base cost, say 10 mils/kWh, the cost of electricity increases very slowly as the uranium price rises. (This is so because fuel is not the main cost of running a nuclear reactor—a typical case reveals 73 percent for capital cost, 5 percent for operation and maintenance, and 22 percent for fuel. Further, the uranium ore cost is not the sole cost of the fuel. It is split roughly equally among ore costs, costs of enriching the uranium, and costs for fabricating the clad fuel rods. At $8/lb, uranium ore costs represent about 7 percent of the final electricity costs.)

It has been argued that with present-day reactors, our uranium reserves will not last the century, and we must therefore push the breeder-reactor program

Table 4.5 Estimate of shale oil resources of world land areas (in billions of barrels).

Continents	Known resources: Recoverable under present conditions	Known resources: Possible extensions of known resources			Marginal and submarginal (oil equivalent in deposits)			Undiscovered and unappraised resources			Order of magnitude of total resources (Oil equivalent in deposits)		
Range in grade (oil yield, in gallons per ton of shale)	10–100	25–100	10–25	5–10	25–100	10–25	5–10	25–100	10–25	5–10	25–100	10–25	5–10
Africa	10	90	Small	Small	ne [a]	ne	ne	4,000	80,000	450,000	4,000	80,000	450,000
Asia	20	70	14	ne	2	3,700	ne	5,400	106,000	586,000	5,500	110,000	590,000
Australia and New Zealand	Small	Small	1	ne	ne	ne	ne	1,000	20,000	100,000	1,000	20,000	100,000
Europe	30	40	6	ne	100	200	ne	1,200	26,000	150,000	1,400	26,000	140,000
North America	80	520	1,600	2,200	900	2,500	4,000	1,500	45,000	254,000	3,000	50,000	260,000
South America	50	Small	750	ne	ne	3,200	4,000	2,000	36,000	206,000	2,000	40,000	210,000
Total	190	720	2,400	2,200	1,000	9,600	8,000	15,000	313,000	1,740,000	17,000	325,000	1,750,000

[a] ne = no estimate.

Source: Duncan and Swanson, *Geological Survey Circular* **523**, 1965, Table 3, p. 18.

Table 4.6   Estimated U.S. uranium resources.

Price ($/lb)	Cumulative resource at or below price (10^4 tons)	Electricity cost due to uranium price (mils/kWh)		Energy content [a] (10^{12} kWh)	
		Water reactor	Breeder	^{235}U	^{238}U
8	59	0.34	0.0024	84	12,000
10	94	0.43	0.0030	130	19,000
15	150	0.65	0.0045	210	30,000
30	220	1.3	0.0090	310	44,000
50	1000	2.2	0.015	1400	200,000
100	2500	4.3	0.030	3500	500,000

[a] For comparison, total U.S. energy use in 1970 was 17×10^{12} kWh.

Source: Uranium resources are as given in USAEC Report WASH–1098 (December 1970), pp. 2–11. Figures for energy content and for electricity cost due to uranium price are by the author. In arriving at electricity cost, I have assumed the energy content of uranium is converted to electricity with an efficiency of 33 percent.

very hard. Resource depletion may occur at prices of $\leqslant$$30/lb, but it certainly will not occur at prices of $\leqslant$$100/lb, unless our total energy use grows preposterously. This very high price for uranium will not quite double the cost of electricity. Further, there is some doubt about the accuracy of Table 4.6 at the higher prices. There is little incentive to prospect for uranium exploitable at $30/lb when the current price is $8/lb. Should the price rise significantly, it would touch off a wave of uranium exploration, which quite possibly would enlarge the resource. Therefore, uranium shortage cannot be used to justify a crash program for the breeder reactor.

On the other hand, it is clear that without the breeding process, present uranium resources won't last indefinitely. As shown in Table 4.6, at $\leqslant$$100/lb the uranium resource can supply 2×10^{12} watts for 200 years. With breeding, these same resources effectively become 100 times larger. Increased uranium costs then contribute negligibly to electricity costs. Even lower-grade ores costing more than $100/lb, could be utilized. Further, thorium becomes a fuel with the advent of breeder reactors. Hubbert gives two examples of the sort of low-grade ores that become usable with breeder reactors—the Chattanooga Shale in eastern Tennessee (0.006 percent uranium, by weight) and the Conway Granite in New Hampshire (0.006 percent thorium, by weight). Since one gram of uranium or thorium, with breeding, has an energy equivalent of 2.7 metric tons of coal, a ton of either of these rocks is equivalent to 160 tons of coal. Needless to say, there is not complete inventory of mineable rock with $\geqslant$ 0.005 percent uranium or thorium content, but such a resource is doubtless very large. With breeder reactors, the fuel resource would essentially be inexhaustible.

Fusion Fuels—Deuterium and Lithium

The physics and technology of fusion energy is discussed in Chapter 5. There it is seen that in fusion reactions using deuterium alone as a fuel, the energy release is 5 to 7 MeV per deuterium atom, or (using 5 MeV) 0.24×10^{12} joules per gram of deuterium consumed. For the fusion reactions using deuterium and lithium-6 jointly as fuel, the energy release is 22.4 MeV per ^{6}Li atom, or 0.36×10^{12} joules per gram of ^{6}Li consumed. It should be stressed that fusion as an energy source is still in an experimental stage, with feasibility far from demonstrated. At this point we cannot say what the efficiency will be; what fraction of the energy release (if any) will be available for use by man. For purposes of discussion here, we will make the optimistic assumption that efficiencies will be comparable to those of currently used devices, such as internal combustion engines, fossil fuel electric generating plants, and so forth. With this assumption, energy release from fusion fuels can be compared directly, joule for joule, with conventional fuels.

The handiest large supply of deuterium is the oceans. The abundance of deuterium in water is one atom of deuterium for every 6500 atoms of hydrogen, or 34 grams deuterium per metric ton of water. With a fusion energy potential of 8×10^{12} joules, a metric ton of water is equivalent to 270 metric tons of coal. And the supply of water considerably exceeds the supply of coal. The oceans contain about 1.5×10^9 cubic kilometers or 1.5×10^{18} metric tons of water. One percent of the deuterium in the oceans has an energy equivalent 500,000 times larger than the world's estimated coal reserves. Deuterium supply can indeed be considered inexhaustible.

The situation with lithium-6 is not so rosy. Presently known lithium resources are estimated at 10^7 tons of natural lithium. The abundance of ^{6}Li is 7.4 percent of natural lithium, so the 10^7 tons convert to 7.4×10^{11} grams ^{6}Li, or 2.7×10^{23} joules energy equivalent. This amount is essentially equal to the world's fossil fuel supply. At present, with no large-scale demand for lithium, the low-grade ores are not well evaluated. The average abundance of lithium in the earth's crust is approximately 25 parts per million, so that one ton of the earth's crust is equivalent to about 20 tons of coal. Whether "average crust" would be a useful energy source would depend on the cost and energy requirements for removing the lithium from the crust. Certainly an ore would not have to be much above average to be useful.

4.5   Commentary

In any discussion of supply of the various fossil fuels, mention should be made of their convertibility. The fluid fuels, oil and natural gas, are most popular today because of ease of transport, handling, and use, and because they either contain fewer polluting impurities or can more readily have them removed. But supplies of oil and natural gas are limited. As these supplies become exhausted, they can be replaced by liquid and gaseous substitutes made from coal, oil shales, or tar sands. The technological feasibility of this has been demonstrated. All that is at issue is the cost, and the cost of oil from coal or gas from coal using currently known technology is not far above the price of the natural products.

Over the long haul of several centuries, only three of the energy sources we've discussed are with certainty large enough to last—uranium and thorium via the breeder reactor, deuterium via fusion, and solar energy. Breeder reactors are technically feasible, but serious engineering problems remain to be solved, and in some respects the whole method is environmentally unsatisfactory (see Chapter 6). Fusion looks environmentally more attractive, but it may in fact never work. Solar energy has been used for some time on a very small scale. It is unclear at this time whether it is technologically feasible on a large scale at a sufficiently low price.

Questions and Problems

4.1
Calculate the energy equivalent in tons of coal, barrels of oil, thousands of cubic feet of natural gas, and kilograms of uranium-235 of the solar energy incident on one square mile of the earth's surface in a 24-hour period.

4.2
The residential power requirement of a typical family of four is 8 kW. (This is total home energy use, not just electrical energy use, and includes the energy wasted in generating and transporting electricity.) Express this requirement in terms of monthly requirements of coal, oil, or natural gas. Also express it in terms of required area for collecting solar energy. How does this area compare with that of a typical house roof?

4.3
Suppose you wish to build a large electric generating plant, which will require fuel at a rate of 2000 MW.

a) Express this requirement in terms of daily requirements of coal, oil, natural gas, and uranium-235. Also express it in terms of required area for collecting solar energy.

b) In the case of fossil fuels, suppose the fuel is being delivered to the firing chamber by a pipe or conveyer, with cross-sectional area of 1 m². In each of the three cases (coal, oil, natural gas), at what speed must the fuel move?

4.4
Natural gas at stp has an energy per unit volume a factor of 1000 lower than that of oil or coal. Is this a significant disadvantage? Support your answer.

4.5
At what velocity does natural gas at stp have the same energy flux as sunlight?

4.6 *
Consider a coastal basin of area A in a region where the tidal range (change in sea level between high tide and low tide) is h.

a) If a dam were built separating the basin from the sea, and 100 percent efficient turbines extracted energy from the flow of water between basin and sea, what is the maximum amount of energy that could be obtained per day? How is this maximum amount of energy distributed over time? What problems are presented by this time distribution?

b) Investigate how the available daily energy is reduced as its time distribution becomes more even. As an example, suppose that when flow is permitted between basin and sea, the flow rate is kept constant. If flow takes place over half of the tidal cycle, how does available energy compare with flow occurring only at low and high tides?

c) Passamaquoddy Bay has an area of 262 km² and a tidal range of 5.5 m. What is the maximum amount of energy per day that could be obtained from this site? Assuming a conversion to electricity with an efficiency of 25 percent of the maximum available energy, what would the daily average power output be? (For comparison, modern fossil fuel or nuclear power plants are typically in the 500 to 1000 MW_e range.)

4.7

a) Derive an expression for the hydroelectric potential of a river. Take the flow rate $Q(z)$ and elevation above sea level $Y(z)$ as known functions of z, the distance upstream from the river mouth.

b) The St. Lawrence River as it leaves Lake Ontario has a flow of 2.4×10^5 ft^3/sec. The elevation of Lake Ontario is 245 ft above sea level. Neglecting the increase in flow downstream of Lake Ontario, calculate the hydroelectric potential of the St. Lawrence River.

4.8

In Section 4.4, the several estimates of stored energy reserves were translated into more readily comprehensible terms by computing how long each source would last if it were supplying all the energy needs at present use levels for the United States, or at twice present use levels for the rest of the world. However, both U.S. use and world energy use have been growing, increasing 3.5 percent per year, a growth rate which will cause them to double in 20 years. Assume that both the current U.S. power use of 2×10^{12} watts and the current world power use of 6×10^{12} watts continue to grow exponentially, at an annual rate of 3.5 percent, indefinitely.

a) Calculate how long each of the following (U.S. and/or world) energy sources would last if it supplied all energy needs: (1) oil and gas, (2) coal, (3) uranium (without breeding), (4) uranium (with breeding), (5) fusion using lithium-6, and (6) fusion using only deuterium.

b) A continuous source of energy can supply energy needs indefinitely, as long as those needs are well below the rate at which the continuous source is naturally dissipated. At present growth rates, in how many years will human power use, now 1/30,000 the total supply of solar energy reaching earth, become larger than the total supply, rendering that source inadequate for filling all our needs?

Comment: From this problem you can see that no energy source is adequate in the face of an exponentially growing demand. In this case, an increase in supply by a factor of 10 adds only 65 years to the lifetime of the source. Moreover, as will be seen in Chapter 6, the natural environment cannot survive exponentially growing human energy use for many more decades. Long before human power use equals solar power levels, environmental problems will become overwhelming. Let us assume (hope) that growth in human energy use moderates and then stops. To accurately evaluate the long-term adequacy of any supply, one must know at what level the energy growth curve levels out. Our lack of knowledge about future demand is much greater than the uncertainties in our knowledge about supply.

4.9 *

a) From the *Handbook of Chemistry and Physics* (or similar reference), obtain the chemical composition and energy content of coal from several different sources. For each variety of coal, estimate the energy content from the information on chemical composition given, and see if your results are in agreement with the energy content value given in the *Handbook*. Obtain an empirical chemical formula $CH_wO_xN_yS_z$ (i.e., determine the decimals w, x, y, and z), for each variety of coal.

b) Repeat the above for crude oil from different sources, and for natural gas from different sources.

c) Compare the empirical formulas obtained with that for living organic matter.

Related Reading

National Academy of Sciences
Resources and Man, Committee on Resources and Man, National Academy of Sciences—National Research Council. San Francisco: W. H. Freeman, 1969. Contains a long, very good chapter by M. K. Hubbert on energy resources.

P. K. Theobald, S. P. Schweinfurth, and D. C. Duncan
"Energy Resources of the United States," USGS Circular 650. Washington, DC: U.S. Geological Survey 1972. Gives various different estimates of U.S. energy reserves.

M. K. Hubbert
"U.S. Energy Resources, A Review as of 1972," a background paper, Committee on Interior and Insular Affairs, U.S. Senate, 93rd Congress, 2nd session. Washington, D.C.: U.S. Government Printing Office, 1974.

5 The Technology of Energy Use

Glendower: I can call spirits from the vasty deep.
Hotspur: Why, so can I, or so can any man;
             But will they come when you do call
             for them?

W. Shakespeare, *King Henry IV*, Part I, Act III

5.0   Introduction

Man uses energy to produce light and heat (for space heating, cooking, water heating, and industrial process heat), and to do work. In obtaining work, the energy often passes through a stage as heat energy. The dominant current devices for delivering mechanical work are the internal combustion engine and the electric motor. The latter brings up the subject of generation of electricity.

From the above we see that the technology of energy use is essentially equivalent to the technology of converting energy from one form to another. There is a very large number of such conversion processes, historical, current, and proposed. The dominant current conversions are (1) chemical energy (fossil fuels) to heat energy for direct use (furnaces and boilers of various sorts), (2) chemical energy to heat energy to mechanical energy (internal combustion engine), (3) chemical to heat to mechanical to electrical energy (gas and steam turbines driving electric generators), (4) gravitational to mechanical to electrical energy (hydroelectric plants), and (5) electrical energy to light (electric lights), electrical energy to heat (ohmic heaters of various sorts), and electrical energy to mechanical energy (electric motors). A conversion process that will soon be of major importance is that from nuclear (uranium) to heat energy (via fission), followed by conversions to mechanical and then electrical energy (nuclear power plants).

The efficiency of energy conversion is a matter of considerable importance. By *efficiency* is meant that fraction of the energy initially present that is converted to the desired end use. For example, in converting chemical energy to heat energy for space heating in a home furnace, the efficiency is defined as the heat delivered throughout the house divided by the energy content of the fuel consumed. Sources of inefficiency are incomplete combustion of fuel and escape of heat up the chimney. There are theoretical upper limits to efficiencies, which derive from the First and Second Laws of Thermodynamics (see Appendix A). Actual efficiencies will always be below these limits, often by a factor of two or more. Some of the conversion processes involve several conversions in series, and inefficiencies will accumulate at each step. A list of some typical efficiencies is given in Fig. 5.1.

There are two reasons for seeking high efficiency, one obvious and one less so. (1) The higher the efficiency, the less fuel is required to accomplish a given task. This results in lower dollar costs for fuel, slower fuel resource depletion, and less environmental damage in the extraction and combustion of the fuel. (2) Since energy is conserved, any energy not used for the desired purpose (due to the efficiency being less than 100 percent) must be disposed of in some other way. The classic example is the waste heat dumped into the condensing water of a steam-electric generating plant. This *thermal pollution* has become a serious environmental problem (see Chapter 6, Part B). The higher the efficiency, the less waste energy to be disposed of.

In the space of one chapter, it is possible to discuss only a small fraction of the energy conversion processes of present or potential future importance. In succeeding sections, we will discuss (1) efficiency of conventional fossil-fuel electric generating plants, (2) nuclear fission reactors for electric power generation, (3) power from nuclear fusion, (4) solar energy schemes, (5) fuel cells, (6) heat pumps, and (7) conversion of solid fuels to fluid fuels. The treatment will be rather general. More detailed treatment and consideration of other conversion devices can be found by pursuing the references at the end of the chapter.

5.1   Fossil-Fuel Fired Electric Generating Plants

These plants, currently responsible for generating the bulk of the electricity in the United States, involve three energy-conversion steps: (1) chemical to thermal, (2) thermal to mechanical, and (3) mechanical to electrical. The first step is accomplished by burning the fuel (coal, oil, or natural gas) in a firing chamber. The heat energy is transferred from the combustion products to water which circulates through tubing surrounding the firing chamber. Inefficiencies in this step

Fig. 5.1 Efficiency of various energy converters. For a device that ▶ involves a sequence of energy conversions, the overall efficiency is indicated. For example, a steam power plant involves conversion of chemical energy to thermal energy, followed by conversion of thermal energy to mechanical energy, followed by conversion of mechanical energy to electrical energy. The overall efficiency of 40 percent is indicated by a bar at 40 percent in all three columns. (From Claude M. Summer, "The Conversion of Energy," *Sci. Amer.* **225**, p. 151. Copyright © 1971 by Scientific American, Inc. All rights reserved.)

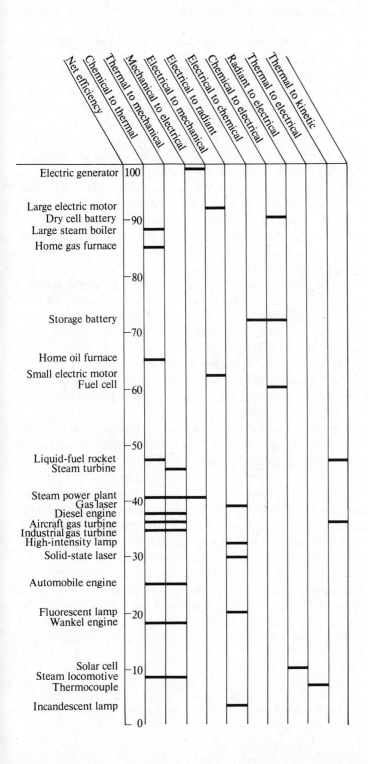

are due to incomplete combustion and loss of heat up the smokestack, and are typically 12 percent.

In step two, the heat energy, now residing in steam at high pressure and temperature, is converted to mechanical energy by allowing the steam to pass through a steam turbine. The principle of the turbine is similar to that of a pin-wheel. A rotating shaft is mounted inside a stationary housing. Appropriately shaped blades project radially outward from the shaft and radially inward from the housing. Steam is introduced at high pressure at one end of the turbine. Driven by the pressure differential, it flows toward the other end of the turbine. It is deflected by the stationary blades on the housing and strikes the blades on the rotating shaft, driving the shaft around. On leaving the turbine, the steam enters a condenser, where it gives up heat and condenses into water. The phase change from gaseous to liquid state causes the specific volume to drop by a large amount, creating a partial vacuum. This maintains the pressure differential across the turbine—high pressure at the entrance and low pressure at the exit. The water from the condenser is then pumped back to the boiler, and the cycle repeats. A large steam turbine-generator is shown in Fig. 5.2.

In the third step, the mechanical energy of the spinning turbine drives a generator, usually coaxial with the turbine. Here electrical conductors attached to a rotor cut across stationary magnetic lines of force, generating electricity by induction. The efficiency of an electric generator is very high—99 percent of the mechanical energy is converted to electrical energy.

An electric generating plant is characterized by the thermal power input (megawatts thermal, MW_t) and by the electrical power output (megawatts electrical, MW_e). $MW_t \times \text{efficiency} = MW_e$. Typical power plants being built during the 1970's range from 500 to 2000 MW_e.

Let's look at the thermodynamics of step two, viewing the process as a heat engine. Figure 5.3 shows the temperature, pressure, and specific volume of the working medium of the engine (water and steam) as it circulates around the system. Water returning from the condenser at a low temperature and pressure passes through some pumps and heaters, so that it enters the boiler at a high temperature and pressure. In the boiler, it absorbs a large amount of heat energy and undergoes a phase change from liquid to gas (water to steam) with little change in temperature. The steam is heated further, and then enters the turbine. (The turbine shown in Fig. 5.3 has three stages—high pressure, intermediate pressure, and low pressure. Each stage is like the turbine described above. As steam passes through the turbine, its volume increases considerably, by a factor of 10^3, so the design details change as the steam pressure falls and the steam volume rises. A better design can be achieved by splitting the turbine into stages.) The steam then undergoes an adiabatic expansion in the high-pressure stage,

Fig. 5.2 Cutaway view of a 550-MW_e steam turbine-generator. The turbine has four stages, ▶ all mounted on the same axis. (The final stage is shown covered by its housing. The generator is in the boxlike enclosure just to the right of the fourth stage of the turbine. (Photo courtesy of General Electric Corporation.)

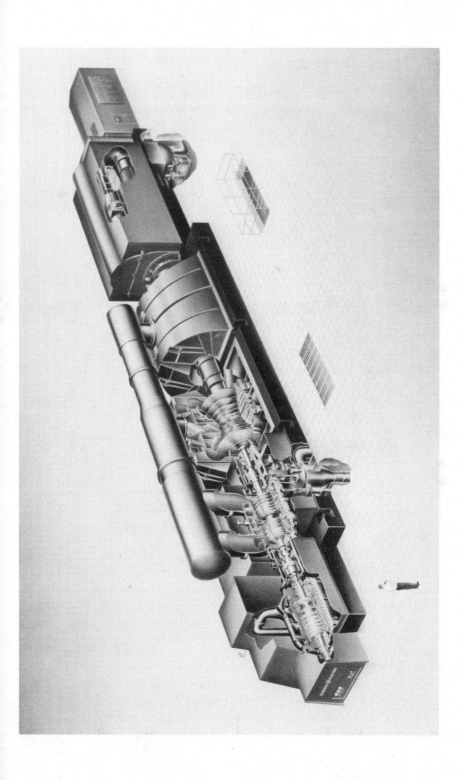

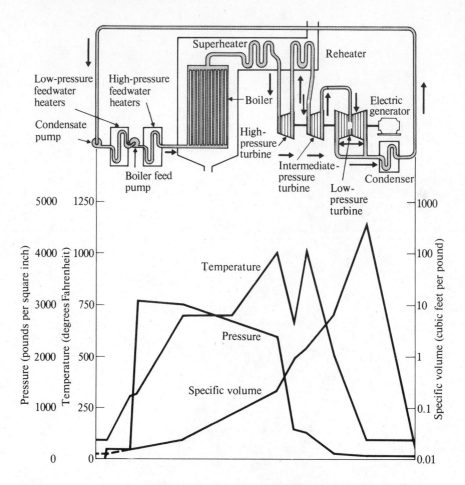

Fig. 5.3 Temperature, pressure, and specific volume of the working medium of a steam turbine. (From Walter Hossli, "Steam Turbines," *Sci. Amer.* **220**, p. 103. Copyright © 1969 by Scientific American, Inc. All rights reserved.)

dropping in pressure and temperature and increasing in specific volume. The steam is reheated and then passes through the intermediate- and low-pressure stages of the turbine, again undergoing adiabatic expansion. It then enters the condenser, where it gives up heat to a low-temperature bath (e.g., water from a nearby lake or river) and condenses, dropping four orders of magnitude in specific volume. It is now ready to repeat the cycle.

Indeed this is a heat engine. Heat is taken from the firing chamber at high temperatures. Some is converted to work (mechanical energy) and the rest is discharged to the condenser at a low temperature. There is a maximum possible efficiency, given by the First and Second Laws of Thermodynamics. This maximum efficiency can be calculated by evaluating how much entropy flows into the working medium along with the inflow of heat. At least this amount of entropy must flow out when heat is discharged, and this will determine the minimum amount of heat to be discarded. Formally,

$$S_{in} = \int \frac{dQ_{in}}{T} \tag{5.1}$$

$$Q_{in} = \int dQ_{in} \tag{5.2}$$

$$S_{out} = \int \frac{dQ_{out}}{T} \tag{5.3}$$

$$Q_{out} = \int dQ_{out} \tag{5.4}$$

$$W = Q_{in} - Q_{out} \tag{5.5}$$

$$S_{out} \gtrless S_{in}, \tag{5.6}$$

where Q_{in} (Q_{out}) is the amount of heat entering (leaving), S_{in} (S_{out}) is the amount of entropy entering (leaving), W is the work done, and T is the temperature at which the heat transfer in question occurs. The outflow of heat occurs at a constant temperature, which is that of the cooling water in the condenser. Call it T_0. The inflow of heat does not occur at a constant temperature. Define an effective temperature for heat inflow T_I^{eff} by

$$T_I^{eff} = \frac{Q_{in}}{S_{in}} = \frac{\int dQ_{in}}{\int \frac{dQ_{in}}{T}}. \tag{5.7}$$

Then

$$S_{in} = \frac{Q_{in}}{T_I^{eff}} \lessgtr S_{out} = \frac{Q_{out}}{T_0}, \tag{5.8}$$

The maximum efficiency will occur when the equality holds, i.e., when

$$Q_{out} = Q_{in} \left(\frac{T_0}{T_I^{eff}} \right). \tag{5.9}$$

The efficiency η is:

$$\eta = \frac{W}{Q_{in}} = \frac{Q_{in} - Q_{out}}{Q_{in}} = 1 - \frac{T_0}{T_I^{eff}}. \tag{5.10}$$

This is exactly the expression for the efficiency of a Carnot cycle, with the temperature of the heat source now given as an effective temperature, which is an appropriate average over the various temperatures at which heat flows into the working medium. For the cycle shown in Fig. 5.3, most of the heat flows in during the phase change from water to steam, at 700°F. Some heat flows at a lower temperature, in the feedwater heaters, and some flows at a higher temperature, in the super heater and reheater. As a reasonable approximation, we would expect $T_I^{eff} \approx 700°F = 644°K$. With a condenser temperature $T_0 = 80°F = 300°K$, this implies a maximum possible efficiency

$$\eta \approx 1 - \frac{300}{644} = 53 \text{ percent.} \tag{5.11}$$

A more careful calculation confirms this figure as the limiting efficiency for the cycle.

In actual practice, a well-designed modern steam turbine achieves 88 percent of the limiting efficiency of 53 percent, or about 47 percent. The overall efficiency of conversion of fossil fuel energy to electrical energy is the product of the efficiencies of the three conversion steps: $0.88 \times 0.47 \times 0.99 = 41$ percent. This overall efficiency is rather low, and it appears that the second step (thermal energy to mechanical energy) is to blame. Yet the turbine design achieved 88 percent of the maximum possible efficiency, which is quite good. Further, on looking into the theoretical maximum efficiency of conversion of the chemical energy of fossil fuels to mechanical energy, one finds it is very high, close to 100 percent. However, in step one when the fossil fuel is burned, there is a large entropy increase, and energy formerly available to do work becomes unavailable. It is thus in step one, rather than step two, that the loss really occurs. In burning the fuel to obtain heat at the "low" temperature of 700 to 1000°F, rather than at the much higher temperatures which can be achieved with fossil fuels, the opportunity to obtain a high efficiency is lost. (Recall the discussion of characteristic temperature of fuel in Section 4.1.) Unfortunately there are materials problems in the firing chamber, boiler, and turbine which currently prevent an increase in the temperature. In a sense, the blame for not being able to obtain a higher overall efficiency is shared by steps one and two.

5.2   Power from Nuclear Fission[1]

Nuclei of intermediate size (neither very small nor very large) are more tightly bound than either of the extreme cases. This means that if a large nucleus breaks into two intermediate-sized pieces, a process called *fission*, energy will be released. However, there is a potential barrier that prevents a nucleus from fissioning—it is necessary to add energy to the nucleus to overcome this barrier. As shown in Fig. B.5, if an energy E_c is added to the nucleus, it will fission, releasing energy E_2. This yields a net energy of $E_2 - E_c$.

One way to add the energy E_c necessary for initiating fission is through *neutron capture:*

$$_Z^A X_N + n \rightarrow {}_{Z}^{A+1} X^*_{N+1}.$$

The excitation energy of the nucleus formed is the sum of the kinetic energy that the neutron had before capture and the energy with which the neutron is bound to the nucleus. Consider neutron capture on the isotopes of ^{235}U and ^{238}U, leading to ^{236}U and ^{239}U. Approximate values of E_c and the binding energy of the captured neutron are shown for the two nuclides:

	E_c	E_{bind}
^{236}U	6.6 MeV	6.8 MeV
^{239}U	7.0 MeV	5.5 MeV

On capture of a neutron with negligible kinetic energy (a "slow neutron"), ^{235}U will fission since $E_{bind} > E_c$. On the other hand, ^{238}U requires capture of a "fast neutron" with kinetic energy $\approx$ 1.5 MeV (experimentally 1.1 MeV) to initiate fission.

The fission process, going from high Z to medium Z, ends up with more neutrons than are appropriate for stability (see Fig. B.3). Invariably a few are "boiled off" in the fission process or emitted immediately by the fission fragments:

$$\text{Uranium} \rightarrow X + Y + 2\text{--}3 \text{ n}. \qquad (5.12)$$

The neutrons released are "fast," typically 1 to 2 MeV. If one can design an assembly of uranium and other materials so that, on the average, one of the neutrons released in a fission process is captured on another uranium nucleus and leads to another fission, which leads in turn to another fission, etc., then a *sustained chain reaction* is taking place:

$$
\begin{aligned}
&\text{n} + \text{U} \rightarrow X + Y + 2\text{--}3 \text{ n} \qquad\qquad\qquad\qquad\qquad (5.13)\\
&\qquad\quad \hookrightarrow \text{n} + \text{U} \rightarrow X' + Y' + 2\text{--}3 \text{ n}\\
&\qquad\qquad\qquad\quad \hookrightarrow \text{n} + \text{U} \rightarrow X'' + Y'' + 2\text{--}3 \text{ n}\\
&\qquad\qquad\qquad\qquad\qquad\qquad\qquad \hookrightarrow \text{etc.}
\end{aligned}
$$

If the situation is such that, on the average, more than one released neutron leads to another fission, then a *divergent chain reaction* is taking place:

$$
\begin{aligned}
&\qquad\qquad\qquad\qquad\qquad\qquad\qquad\qquad\qquad \rightarrow \text{etc.}\\
&\qquad\qquad\qquad\qquad \rightarrow \text{n} + \text{U} \rightarrow X^{3'} + Y^{3'} + 2\text{--}3 \text{ n}\\
&\qquad\qquad\qquad\qquad\qquad\qquad\qquad\qquad\qquad \hookrightarrow \text{etc.}\\
&\qquad\quad \rightarrow \text{n} + \text{U} \rightarrow X' + Y' + 2\text{--}3 \text{ n}\\
&\qquad\qquad\qquad\qquad\qquad\qquad\qquad\qquad\qquad \rightarrow \text{etc.}\\
&\qquad\qquad\qquad\qquad \hookrightarrow \text{n} + \text{U} \rightarrow X^{4'} + Y^{4'} + 2\text{--}3 \text{ n}\\
&\qquad\qquad\qquad\qquad\qquad\qquad\qquad\qquad\qquad \hookrightarrow \text{etc.}\\
&\text{n} + \text{U} \rightarrow X + Y + 2\text{--}3 \text{ n} \qquad\qquad\qquad\qquad\qquad (5.14)\\
&\qquad\qquad\qquad\qquad\qquad\qquad\qquad\qquad\qquad \rightarrow \text{etc.}\\
&\qquad\qquad\qquad\qquad \rightarrow \text{n} + \text{U} \rightarrow X^{5'} + Y^{5'} + 2\text{--}3 \text{ n}\\
&\qquad\qquad\qquad\qquad\qquad\qquad\qquad\qquad\qquad \hookrightarrow \text{etc.}\\
&\qquad\quad \hookrightarrow \text{n} + \text{U} \rightarrow X'' + Y'' + 2\text{--}3 \text{ n}\\
&\qquad\qquad\qquad\qquad\qquad\qquad\qquad\qquad\qquad \rightarrow \text{etc.}\\
&\qquad\qquad\qquad\qquad \hookrightarrow \text{n} + \text{U} \rightarrow X^{6'} + Y^{6'} + 2\text{--}3 \text{ n}\\
&\qquad\qquad\qquad\qquad\qquad\qquad\qquad\qquad\qquad \hookrightarrow \text{etc.}
\end{aligned}
$$

The chain reaction is the concept basic to both nuclear reactors and nuclear bombs. In a reactor, a sustained chain reaction occurs, and the rate at which fission takes place remains constant. In a bomb, a divergent chain reaction occurs, and the rate at which fission takes place grows rapidly with time.

We define the *effective reproduction constant* k_e as the average number of neutrons from one fission that go on to cause fissions in the next "generation." If k_e is less than one, the number of fissions per generation will shrink with time, and no chain reaction will occur. If k_e is equal to one, the number of fissions per generation remains constant, and a sustained chain reaction can take place. If k_e is greater than one, the number of fissions per generation increases with time, and a divergent chain reaction takes place. Since the average fission releases about 2½ neutrons, k_e could be as large as 2.5 if all the neutrons were used. However, neutrons are wasted in the following two ways. (1) They leak out the edges of the assembly. This loss can be reduced by making the assembly larger (more inside volume per unit surface area). (2) They are captured (by fuel or other parts of the reactor) in reactions not leading to fission.

Consider the fate of fission neutrons in a large piece of natural uranium (0.7 percent ^{235}U and 99.3 percent ^{238}U). Neutrons of 1–2 MeV predominantly interact with ^{238}U by *inelastic scattering*:

$$n + {}^{238}U \rightarrow n' + {}^{238}U^*, \qquad {}^{238}U^* \rightarrow {}^{238}U + \gamma.$$

The neutron gives up energy, exciting the nucleus, which subsequently decays by gamma emission. In this way, the neutrons lose energy, quickly dropping below 1 MeV, that is, below the ^{238}U fission threshold. Once the neutrons have less energy than is necessary to excite ^{238}U to its first excited state, inelastic scattering can no longer take place. The neutrons then lose energy, considerably more —slowly, by *elastic scattering*—billiard ball-like collisions with uranium or other nuclei in which the neutrons impart a small amount of kinetic energy to the struck nucleus, losing a small amount of energy with each collision. As the neutrons are slowed down toward thermal energies ($\sim 1/40$ eV), they pass through an energy region of *resonance capture* (1–100 eV), a region in which there are narrow energy bands where the reaction $n + {}^{238}U \rightarrow {}^{239}U^* \rightarrow {}^{239}U + \gamma$ occurs with very high probability. Unless neutrons pass through the resonance capture region rapidly, they will be lost. Since neutrons in a piece of natural uranium lose energy slowly in the resonance capture region, a simple large block of natural uranium will not give rise to chain reactions, i.e., its value of k_e is less than one.

Two schemes suggest themselves to circumvent the difficulties outlined above. (1) By using a uranium fuel highly enriched in ^{235}U, the amount of neutron inelastic scattering on ^{238}U is reduced. A larger fraction of the fast neutrons is captured by ^{235}U, leading to fission. Note that in this reactor concept, it is the fast neutrons that are being utilized. (2) Alternatively, one can use thermal neutrons for initiating fission, utilizing an arrangement that causes the neutrons to lose energy rapidly as they pass through the resonance capture region. This arrangement involves a *moderator,* a material of low atomic mass number (A)

interposed between pieces of fuel. When a neutron scatters elastically from a nucleus, the amount of energy lost depends on the mass of the nucleus. The lower the mass, the more energy a neutron will lose in scattering from it. Neutrons lose considerably more energy in scattering elastically from protons, deuterons, or even carbon nuclei than they do in scattering elastically from uranium nuclei. In addition to having low A, a moderator must have a low capture cross-section for neutrons. Otherwise the moderator, in reducing one source of neutron loss (resonance capture on ^{238}U), will introduce another one (capture on the moderator nuclei). Water (H_2O), heavy water (D_2O), and carbon have been used as moderators. D_2O and carbon allow natural uranium to be used as a fuel. With H_2O, the reaction $n + p \rightarrow d + \gamma$ taking place on the hydrogen nuclei causes enough of a neutron loss that k_e is less than one if natural uranium is used. By using uranium slightly enriched in ^{235}U, values of k_e greater than one can be achieved.

If k_e is not exactly one, the power level in a reactor will grow or shrink. We must now consider the question of stability; how to keep k_e exactly unity. First consider the time scale involved. Neutrons take typically 0.1 to 1.0 milliseconds to slow down from fission neutron energies (1–2 MeV) to thermal energies (1/40 eV) and be captured, initiating the next generation of fissions. In a fast reactor, the time from one generation to the next is much shorter, typically 10^{-7} seconds. However, in both types of reactors a small fraction of the neutrons (about 0.65 percent) are released not in the fission process, but in the radioactive decay of fission products, e.g.,

$$^{87}Br \xrightarrow{\beta^-, \ 56 \ sec} {}^{87}Kr^* \longrightarrow {}^{86}Kr + n.$$

These *delayed neutrons* have an average delay of 14 seconds.

Consider how a reactor would behave if the delayed neutrons were not present. Suppose a reactor were adjusted so that $k_e = 1.001$. Power levels would increase by 0.1 percent per generation, doubling in 700 generations—0.07 to 0.7 sec for a thermal neutron reactor, and ~10^{-4} sec for a fast-neutron reactor. These times are rather short, particularly the fast-reactor case. Little time would be available for any adjustments. The delayed neutrons have the effect of lengthening the time between generations to (10^{-7} sec or 10^{-3} sec) + 0.0065 × 14 sec ≈ 0.1 sec, for both slow and fast reactors. With this longer generation time, if $k_e \approx 1.001$, it takes 70 seconds for the power level to double. Adjustments and corrections can be made in a more leisurely fashion. Note, however, if k_e becomes sufficiently large, then the chain reaction will take place independently of the delayed neutrons, and the shorter time scale applies.

There are three feedback mechanisms that tend to hold k_e at unity. (1) Use of *control rods*. These rods, made of material that strongly absorbs neutrons, can be inserted into or withdrawn from the reactor, lowering or raising k_e. This is the external control mechanism used to start up, adjust, and shut down a reactor. The feedback loop is "man-made" rather than innate, and in principle could malfunction. (2) Doppler broadening of the ^{238}U resonance capture cross sections.

These cross sections are inherently very narrow, and neutrons are absorbed only if they have precisely the right energy (in the rest frame of the ^{238}U nucleus). If the ^{238}U nucleus is moving, the effective neutron energy is Doppler shifted. If the ^{238}U is hot, the nuclei are in random thermal motion and the resonance absorption line is broadened. Neutrons are thus more easily absorbed. The hotter the fuel, the more the resonances are broadened and the more readily they capture neutrons. If $k_e > 1$, the power level in the reactor will rise, heating the fuel. Doppler broadening and neutron capture rates will increase, and k_e will fall. Similarly if $k_e < 1$, the power level will fall, the fuel will cool, the resonances will narrow, capture rates will drop, and k_e will then rise. This negative feedback mechanism tends to hold k_e equal to one. (3) Temperature variation of the density of the moderator. If $k_e > 1$, the power level will rise, the moderator will become hotter and therefore less dense. It will moderate less effectively and more neutrons will be lost, causing k_e to fall. For $k_e < 1$, the inverse occurs. Again, we have a negative feedback mechanism which tends to hold k_e equal to one.

The energy of fission is released dominantly in the form of kinetic energy of the fission fragments. The *energy of the radioactive decay* of the fission fragments is not negligible, however, being about 8 percent of the total energy release. In both cases, the energy is quickly transformed to heat energy of the fuel element. Thus most of the heating occurs at the instant of fission, but 8 percent occurs later as the fission products decay. Because some of the decay times are appreciable (minutes to days), a reactor will continue generating heat after it has been turned off, i.e., after the fission chain reaction has been stopped. The inability to instantly stop the energy release in a nuclear reactor causes safety problems (see Chapter 6, Part C).

In the United States, most reactors used for electric power generation are water-moderated slow-neutron reactors—either pressurized-water reactors (PWR) or boiling-water reactors (BWR). Schematics of both are shown in Fig. 5.4. The fuel elements, uranium oxide slightly enriched in ^{235}U, are immersed in water, which both moderates the neutrons and carries the heat away from the fuel elements. High-temperature, high-pressure steam is generated in the reactor vessel of the BWR. In other words, the moderator (water) boils. In the PWR, pressure and temperature are arranged so that the moderator does not boil. Rather it passes through a heat exchanger, transferring its heat to water in a separate circuit which does boil, creating steam. A cutaway view of a PWR and associated piping, pumps, and steam generators is shown in Fig. 5.5. From this point on, the plant is similar to a fossil-fuel plant. The steam drives a steam turbine, which in turn drives a generator. Because of materials limitations in the reactor, the temperature of the steam from a nuclear reactor is lower than that from a fossil-fuel plant, and so the efficiency is lower (33 percent rather than 40 percent).

Breeding

Uranium-235 is a rare (0.7 percent) isotope of a rare ($8/lb) element. From the binding energy *vs. A* curve (see Fig. B.2), we see that all nuclides at the

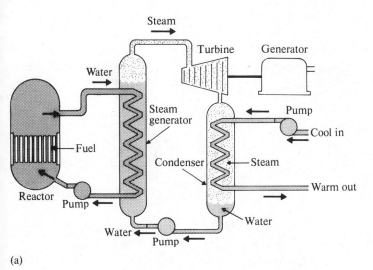

(a)

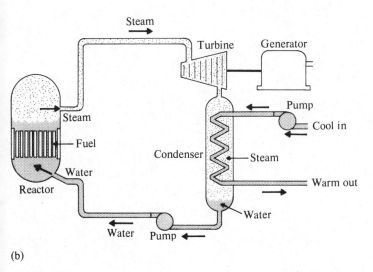

(b)

Fig. 5.4 (a) A pressurized-water reactor (PWR) with heat exchanger (left), supplying steam to the turbine with its steam condenser (right). (b) A boiling-water reactor (BWR). (From David R. Inglis, *Nuclear Energy: Its Physics and Its Social Challenge.* Reading, Mass.: Addison-Wesley, 1973, pp. 96, 97. By permission.

high A end can in principle serve as fuel, if some trick can be found to make them fission. The trick for ^{235}U is neutron capture. Another trick will work for ^{238}U and ^{232}Th—breeding. Breeding requires two (or more) neutrons per fission rather than just one. One neutron is used to keep the fission chain reaction going;

Fig. 5.5 Cutaway view of a nuclear-steam supply system based on a pressurized-water reactor. The reactor is lower center. Driven by pumps (small cylindrical objects mounted on piping), water circulates between the reactor and two steam generators (large cylindrical objects to the right and left) through piping. For scale, note man at far right. (Photo courtesy of Combustion Engineering, Inc. and the U.S. Atomic Energy Commission.)

a second is used to "replace" the used-up nucleus that has just fissioned, and additional neutrons can be used to increase the supply of fissionable fuel.

$$n + {}^{235}_{92}U_{146} \to X + Y + 2\text{–}3\,n \begin{cases} \to n + {}^{235}_{92}U_{146} \to X' + Y' + 2\text{–}3\,n \\ \\ \to \text{Neutron-induced breeding reaction} \end{cases} \tag{5.15}$$

$$n + {}^{238}_{92}U_{146} \to {}^{239}_{92}U_{147} \xrightarrow{\beta^-,\ 20\ \min} {}^{239}_{93}Np_{146} \xrightarrow{\beta^-,\ 2\ \text{days}} {}^{239}_{94}Pu_{145} \tag{5.16}$$

Uranium-238 is converted into Plutonium-239 by neutron capture followed by two β-decays. In a reactor, ${}^{239}Pu$ can be used much as ${}^{235}U$, i.e., the reaction $n + {}^{239}_{94}Pu_{145} \to X' + Y' + 2 - 3\,n$ proceeds very similarly to neutron-capture-induced ${}^{235}U$ fission. In Eq. 5.15 above, ${}^{239}Pu$ can be substituted for ${}^{235}U$. In that arrangement, we see that each time a ${}^{239}Pu$ nucleus fissions, it is replaced by at

Fig. 5.6 A 550-MW$_e$ reactor being loaded with fuel. (Photo courtesy of Iowa Electric Light and Power Company and the U.S. Atomic Energy Commission.)

least one additional ^{239}Pu nucleus bred from ^{238}U. Therefore, ^{239}Pu is not used up. In fact, if the average number of breedings per fission exceeds one, the amount of ^{239}Pu will increase. What is used up is ^{238}U, thus making ^{238}U a useful fuel. As shown in Table 4.6, the breeder greatly increases nuclear energy reserves.

A similar breeding reaction exists for thorium-232:

$$n + {}^{232}_{90}\text{Th} \rightarrow {}^{233}_{90}\text{Th} \xrightarrow{\beta^-} {}^{233}_{91}\text{Pa} \xrightarrow{\beta^-} {}^{233}_{92}\text{U}. \tag{5.17}$$

Like ^{235}U and ^{239}Pu, ^{233}U can be used in fast or slow neutron-induced fission chain reactions.

The breeder reactor is still in a developmental stage, and there is more than one scheme for "how to do it." The concept receiving the most attention is the liquid-metal fast breeder reactor (LMFBR). This reactor works with fast neutrons. As noted earlier, a fast-neutron reactor requires a fuel highly enriched ($\sim$15 percent) in ^{235}U or ^{239}Pu. (Both ^{235}U and ^{239}Pu fission readily with either fast or slow neutrons.) Since no moderator is wanted, water is inappropriate for carrying the heat away from the fuel elements. Sodium metal, a liquid at reactor temperatures, is used instead. The reactor consists of a core region containing the enriched fuel (85 percent ^{238}U and 15 percent ^{239}Pu), surrounded by a blanket region containing natural uranium or even uranium depleted of ^{235}U. The fuel (both in the core and blanket regions) is contained in long, slender, stainless-steel-clad tubes of 0.25-inch diameter immersed in liquid sodium. The heat carried away by the rapidly flowing liquid sodium is used to generate steam, which drives turbines. Because use of liquid sodium does not entail high pressures, the reactor materials can withstand higher temperatures than in an LWR, and so generating efficiencies of 40 percent can be expected.

As a simplified explanation, ^{239}Pu in the core maintains the chain reaction, and excess neutrons leak from the core into the blanket where they are captured by ^{238}U nuclei, thus breeding more ^{239}Pu (see Eq. 5.16). In actual practice, substantial amounts of ^{239}Pu are bred from the ^{238}U in the core. Also, since ^{238}U fissions under fast neutrons, some part of the chain reaction is maintained by ^{238}U fission both in the core and in the blanket. The design is such that for each ^{239}Pu nucleus that fissions, at least one ^{238}U nucleus is converted to ^{239}Pu. Periodically, the fuel elements in both core and blanket must be removed and reprocessed, since (1) the amount of ^{239}Pu in the blanket is building up, (2) the amount of ^{239}Pu in the core is being depleted, (3) the amount of ^{238}U is being depleted in both places, and (4) the concentration of fission fragments is building up. ^{239}Pu is recovered from both the blanket and core elements and is used for making new core fuel elements. ^{238}U is also recovered. Fission fragments are stored in perpetuity, as with a nonbreeder reactor.

The LMFBR lacks some of the inherent stability of the LWR. First, a fast reactor, with its very short time between generations, is vitally dependent on the delayed neutrons for safe operation. If k_e becomes sufficiently large that the reactor is critical without requiring the delayed neutrons, it will rapidly run out of control. However, the fraction of delayed neutrons from ^{239}Pu is only 0.3 percent, compared with 0.65 percent from ^{235}U. This means that a ^{239}Pu-based fast-neutron reactor cannot tolerate as large excursions of k_e above one as can a ^{235}U-based slow-neutron reactor.

Second, of the two inherent negative feedback mechanisms that hold k_e at one in an LWR, one becomes a positive feedback mechanism for the LMFBR. (1) The negative feedback supplied by variation of moderator density with temperature in an LWR is replaced by positive feedback caused by variation of sodium density with temperature in an LMFBR. Specifically, if a region in the core overheats, causing a reduction in liquid sodium density, the reactivity of that region will go up (since the neutrons normally captured by the sodium are now available for inducing fission), causing further overheating. (2) Doppler broadening works in an LMFBR in qualitatively the same way as in an LWR. Higher temperature increases neutron loss through capture.

5.3 Power from Nuclear Fusion

The energetics of fusion follow qualitatively from the low-A end of the binding energy *vs.* A curve (Fig. B.2). As A increases, nucleons become more tightly bound, so energy can be released by two very small nuclei combining to form a single larger nucleus. If one looks at the details of the binding energy curve (not shown), one observes departures from the smooth curve at low A. ^{4}He is anomolously tightly bound; d is very loosely bound. Combining deuterons to make ^{4}He would release a lot of energy. However, one must consider more than just energetics. One must consider which reactions "go" with any appreciable probability. Below is a list of reactions important for fusion power.

$$_1^2d_1 + {}_1^2d_1 \rightarrow {}_2^3He_1 + n + 3.2 \text{ MeV} \qquad (5.18)$$

$$_1^2d_1 + {}_1^2d_1 \rightarrow {}_1^3t_2 + p + 4.0 \text{ MeV} \qquad (5.19)$$

$$_1^2d_1 + {}_1^3t_2 \rightarrow {}_2^4He_2 + n + 17.6 \text{ MeV} \qquad (5.20)$$

$$_1^2d_1 + {}_2^3He_1 \rightarrow {}_2^4He_2 + p + 18.3 \text{ MeV} \qquad (5.21)$$

$$n + {}_3^6Li_3 \rightarrow {}_1^3t_2 + {}_2^4He_2 + 4.8 \text{ MeV}. \qquad (5.22)$$

Reactions 5.18 and 5.19 are about equally probable. One can conceive of a situation in which deuterium is supplied as the fuel, reactions 5.18 and 5.19 proceed, and the tritium and helium-3 produced combine with the fuel via reactions 5.20 and 5.21. The overall equation is

$$6_1^2d_1 \rightarrow 2p + 2n + 2_2^4He_2 + 43.1 \text{ MeV}. \qquad (5.23)$$

The energy release is about 7 MeV per deuteron. If the ^{3}He is not "burned," the reaction becomes

$$5_1^2d_1 \rightarrow p + 2n + {}_2^3He_1 + {}_2^4He_2 + 24.8 \text{ MeV}, \qquad (5.24)$$

and the energy release is about 5 MeV per deuteron. An alternate situation might use deuterium and lithium-6 jointly as a fuel, proceeding via reactions 5.20 and 5.22. The neutrons produced in reaction 5.20 drive reaction 5.22, and the tritons produced in reaction 5.22 drive reaction 5.20. The overall equation is

$$_1^2d_1 + {}_3^6Li_3 \rightarrow 2_2^4He_2 + 22.4 \text{ MeV}. \qquad (5.25)$$

For any nuclear reaction to proceed, the nuclei must be brought close together so that the nucleons of one nucleus "feel" the (short-range) nuclear forces from the other nucleus. But in all proposed schemes, at least one reaction in the chain involves an initial state of two positively charged nuclei (e.g., d + d; d + t; d + ^{3}He). The repulsive coulomb force holds the nuclei apart and prevents fusion reactions from taking place spontaneously on earth—a good thing! To obtain a fusion reaction, the initial-state nuclei must be given sufficient energy to overcome (or penetrate) this coulomb barrier. A kinetic energy in the range of tens of kilovolts is required. Assuming this energy is that due to the random thermal motion of the nuclei, they must be heated to temperatures in the range of 10^8 °K.

The necessary steps in obtaining power from nuclear fusion are the following.

1. Heating the fuel (deuterium, tritium, and/or helium-3 in some combination) to a temperature in the range of 10^8 °K.

2. Confining the fuel (holding it together) for a long enough time so that the number of fusion reactions taking place release more energy than was required for the initial heating.

3. Converting the energy released to some usable form.

At the present moment, these three steps have been accomplished in one device —the hydrogen bomb. Needless to say, the uses to which this energy release can be put are rather limited. Attempts to obtain *controlled* fusion power have met with considerably less success.

The major stumbling block to controlled fusion power so far has been step 2—the confinement problem. The energy required to heat a unit volume of fuel is proportional to the density n of the fuel. But the power produced per unit volume of fuel is proportional to the square of the fuel density, since the probability of collision is proportional to n^2. If the heated fuel is held together for a time τ (the confinement time), then the requirement that the energy released exceed the energy needed for heating can be written as

$$an^2\tau > bn, \qquad \text{or} \qquad n\tau > \frac{b}{a}, \tag{5.26}$$

where a and b are constants of proportionality which can be evaluated. This leads to the Lawson Criterion,

$$n\tau > 10^{14} \text{ sec-particles cm}^{-3}. \tag{5.27}$$

The minimum required confinement time is thus inversely proportional to the fuel particle density.

To shorten the necessary confinement time, one would like to use as high a fuel-particle density as possible. However, the power produced per unit volume goes up as the square of the density. If the power density becomes too high, the pressure becomes too high to allow confinement and it is not possible to extract the liberated energy in a continuous, steady-state fashion. A fuel density of 3×10^{14} particles/cm³ (10^{-5} atm) yields a power density in the tens of watts/cm³ range, which is about the limit. Thus for a steady-state device, a confinement time in excess of 0.1 to 1.0 second is necessary.

Alternatively, one can consider a pulsed device using essentially a series of "microexplosions." Here the instantaneous power density can be much higher, and so the fuel density can be much higher, and the confinement time can thus be much shorter.

Most of the research effort over the past two decades has been directed toward steady-state devices. At a temperature of 10^8 °K, the fuel is totally ionized, i.e., it is a *plasma*. The problem then is to confine a fully ionized gas of approximate density 3×10^{14} particles/cm³ for at least 0.3 second. The often-attempted solution has been *magnetic confinement,* a method using a magnetic field to control the motion of the individual charged particles of the plasma, thus insulating the plasma from the material walls of its container. Success has been limited. In addition to the externally established magnetic field, there are fields that arise due to the plasma itself, due to the collective motion of the ions. This situation can give rise to a host of instabilities which allow the plasma to escape. Over the past two decades, a great deal has been learned about the behavior of plasmas and about plasma instabilities. This knowledge, in turn, has led to longer confinement times. At present, $n\tau$ values are still two orders of magnitude below the Lawson Criterion, but there is optimism in some quarters that devices currently under consideration will meet the Lawson Criterion.

In recent years, a new approach called *laser fusion* has received considerable attention. In this scheme, a tiny pellet (composed of frozen deuterium and tritium) would be compressed by a short ($\sim$10 nsec), high-energy ($\sim 10^5$ joule) pulse of laser light. The light pulse, split and focussed so that it impinges on the pellet from all sides, would isoentropically compress the pellet by a factor of 10^4. The compression process would increase the temperature of the center of the pellet to the 10^8-°K range, initiating fusion reactions. The fusion reactions would further heat the pellet, which would "burn" from the center out in approximately 10^{-11} second, releasing about 10^7 joules of fusion energy. This scheme shows how the microexplosion approach differs radically from the steady-state approach. Fuel densities and power densities would be exceedingly high. Necessary confinement times would be very short (10^{-11} sec after compression) and inertia should suffice to achieve them.

The above scenario has been borne out by computer studies, but not yet by actual experiments. There may be surprises awaiting as one tries to compress solid matter to densities 10^4 times larger than normal. There is also the problem of developing lasers with sufficiently high power capabilities and sufficiently high efficiencies. It is necessary that the useful energy extracted from the microexplosion be more than enough to power the laser. A study that assumes the laser converts the input energy to light with 10 percent efficiency and the fusion energy is converted to electricity with 40 percent efficiency finds that perhaps 30 percent of the electricity produced will be required to power the laser, with 70 percent available for "sale."

A few words are in order about step 3—converting the energy released by fusion to a usable form, in particular to electricity. The released energy initially will be in the form of high energy charged particles and neutrons. These particles can be absorbed in bulk matter, generating heat, which can be used as in a conventional steam-electric generating plant. Higher efficiency may be achievable by converting the charged-particle energy directly to electricity by means of magnetohydrodynamic plasma conversion (see Related Readings).

Research on controlled fusion power has been in progress for two decades, and much has been learned. But the feasibility of fusion power has *not yet been demonstrated.* It is not clear when the proof of feasibility will come, or even whether it will ever come. If feasibility is proved, there will still remain a host of engineering problems to be solved before a functioning pilot plant becomes a reality. And finally, dollar costs and environmental hazards cannot be assessed until full-scale power plants have been designed.

5.4   Solar Energy Schemes

As noted in Section 4.3, solar energy is characterized by a reasonably high quality and an abundant total quantity, but also by a very low concentration. Further, its delivery is not continuous due to nights and cloudy days. These adverse characteristics define the major technological problems of solar energy use—energy collection and concentration, to compensate for the low initial concentration, and energy storage, to compensate for the noncontinuous delivery. Before looking at conceived schemes for using solar energy, it may be useful to reread the discussion of photosynthesis in Chapter 3 to see how nature handles the matter. The fact that nature achieves a solar-to-chemical energy conversion efficiency of only a few percent under favorable conditions should guide our thinking about what man might hope to achieve.

Solar energy schemes can be classified as either *direct*-utilization or *indirect*-utilization schemes. The indirect-utilization schemes allow some natural process to collect the solar energy, concentrate it to some degree, and perhaps also store it. Wind power, water power, and photosynthesis are indirect utilization schemes and were discussed in Section 4.3. Another indirect utilization scheme currently under consideration[2] takes advantage of the temperature difference between the surface water and deep water of the ocean to drive a heat engine. The following discussion will be limited to direct utilization schemes.

The major end uses contemplated for solar energy include home heating, hot-water heating, home cooling, and electricity generation. In most cases, heat energy is either the desired end form or an intermediate stage. To be useful, the heat must be at a temperature above ambient, often considerably above ambient. Since solar energy has a high characteristic temperature, this is in principle possible to achieve. The task of the *collector* is to convert solar radiation to heat at a high temperature with high efficiency. This is accomplished by maximizing the absorbed solar energy and minimizing all energy losses. Consider the idealized situation in which conductive and convective losses are negligible. Then the temperature is determined by the radiative losses—the temperature of the collector rises until the radiative losses equal the difference between the energy flow in and the energy flow drawn from the collector. Minimizing radiative losses is therefore an important consideration in collector design.

Schemes for reducing radiative losses make use of the difference between the spectrum of the incident solar energy (mainly the visible and near infrared) and the spectrum of the reradiated energy (dependent on temperature; in the far infrared for most cases of interest). There is little overlap between these spectra for collector temperatures below 900°K. The standard scheme is to cover the collector with one or more layers of glass or some other material that is trans-

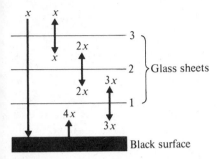

Figure 5.7

parent to solar radiation but black to far infrared. In this way, the solar radiation can enter easily, but the infrared reradiation has difficulty escaping.

A situation using three layers of glass is shown in Fig. 5.7. The collector proper is assumed to be black to both visible and infrared radiation. The incident solar radiation passes through all layers of glass and is absorbed by the collector, which radiates an amount of IR radiation appropriate to its temperature. This IR radiation is absorbed by the first layer of glass, which reradiates an appropriate amount back to the collector as well as out to the second layer of glass. The second layer reradiates in the same way, as does the third. For the case shown in Fig. 5.7, no useful power is being drawn from the collector, so the energy flux going out must equal the energy flux coming in. This condition, flux out equaling flux in, must hold for any surface we wish to consider. Using this fact as well as the fact that each hot glass layer will radiate equal amounts in both directions, we are able to determine all fluxes. If the incident solar flux is denoted by x, then the outermost glass layer (layer 3) must reradiate a flux x outward (and hence also x inward). The inward-going flux between layers 3 and 2 is $2x$ (x solar + x IR from layer 3). Therefore, the outward-going flux (which all comes from layer 2) is also $2x$. Continuing inward, we obtain the fluxes indicated in Fig. 5.7. The collector surface radiates an IR flux of $4x$; in the absence of glass layers, it would radiate a flux x. Therefore, the effect of the covers has been to raise the temperature of the collector by a factor $4^{1/4}$. Expressed more generally, with n layers of glass, incident solar flux S, collector area A, an ambient temperature T_0 surrounding the collector, and a flux q_0 being drawn from the collector, one can show that

$$(n+1)(AS - q_0) = \sigma A(T^4 - T_0^4). \tag{5.28}$$

But $q_0/AS \equiv \eta$, the efficiency with which solar energy is converted to heat and supplied by the collector. So

$$T^4 - T_0^4 = (n+1)\frac{S}{\sigma}(1-\eta). \tag{5.29}$$

Note that as one draws more heat from the collector, increasing η, the temperature falls. This means that in actual use, one must compromise between high temperature and high efficiency.

Equation 5.29 suggests that the temperature of the collector will increase with increasing n roughly as $(n + 1)^{1/4}$. Since it is *in principle* impossible for the collector to attain a higher temperature than the sun (Second Law—heat won't flow from cooler to hotter), the equation must break down at some point. It does when the collector temperature becomes so high that its reradiation is no longer only in the far IR, but also partly in the visible. Once the spectra from the sun and the collector overlap, it is no longer possible to have a material transparent to radiation from one and at the same time black to radiation from the other. (In practice, other considerations limit the temperature of the collector long before it starts radiating appreciably in the visible. The layers of glass will reflect and absorb some of the incident solar radiation, so the number of layers cannot be made too high. Typically, one to four layers are used.)

As an alternative to covering the collector with layers of IR-absorbing material, one can use a collector surface that is highly reflective to infrared radiation, while at the same time is black to visible light. (One method for making such a surface is to coat a shiny metallic surface with a layer of absorbing material a few microns thick, i.e., more than the wavelength of visible light but less than the wavelength of the far infrared. The short-wavelength radiation will be absorbed by the coating, but the long wavelength radiation won't "see" it and will be reflected by the underlying shiny surface.) If a surface reflects IR well, it radiates it poorly, which is the desired property. Let a_V and ϵ_{IR} be the absorption of visible radiation and the emissivity for infrared, respectively. Then with no useful energy being drawn from the collector, the energy flux balance reads

$$Sa_V = \sigma \epsilon_{IR}(T^4 - T_0^4). \tag{5.30}$$

Because ϵ_{IR} has been made small, T^4 must rise for the collector surface to radiate away the same amount of energy. With useful energy being withdrawn from the collector, the equation becomes

$$Sa_V - \frac{q_0}{A} = \sigma \epsilon_{IR}(T^4 - T_0^4), \tag{5.31}$$

$$T^4 - T_0^4 = \left(\frac{S}{\sigma} \frac{a_V}{\epsilon_{IR}} (1 - \frac{\eta}{a_V}) \right). \tag{5.32}$$

The temperature has been increased by a factor of approximately $(a_V/\epsilon_{IR})^{1/4}$. In practice, surfaces with $a_V > 0.9$ and $\epsilon_{IR} \approx 0.1$ can be achieved. While this approach is more effective than using multiple layers of glass, it is more expensive.

A typical *flat-plate collector* consists of a flat surface black to visible light, covered with two to three layers of glass and insulated so as to minimize conductive and convective heat losses. Heat is extracted by circulating water or air through tubing in thermal contact with the plate. The orientation of the collector can be horizontal, vertical facing south (in the northern hemisphere), or preferably, tilted toward the equator through an angle equal to the latitude of the site.

Under favorable conditions, temperatures approaching 100°C can be achieved. Using a selectively absorbing collector ($a_V/\epsilon_{IR} \approx 10$), temperatures approaching 200°C can be achieved.

Using a *focussing collector,* one can achieve considerably higher temperatures. Solar furnaces using high-quality large parabolic mirrors have attained temperatures in excess of 3000°C. However, two considerations argue against focussing collectors unless high temperatures are essential. (1) A focussing collector, with moving parts so that the sun can be tracked, is more complex and more expensive than a flat-plate collector covering the same area. (2) A focussing collector makes use of only the direct light from the sun and not light scattered by clouds or haze. On a cloudy-bright day, a focussing collector is useless, while a flat-plate collector works satisfactorily. (Recall the discussion of photosynthesis in Section 3.1.)

Most schemes for solar energy utilization require provision for storing thermal energy. Energy can be stored as sensible heat (increased temperature of some material) or as latent heat (phase change of some material). Water, with its high specific heat and low cost, is the favored material for sensible-heat storage at temperatures below its boiling point. If air is used as the heat-transfer medium, crushed rock provides the advantage of a large, inexpensive surface area for heat transfer, but has only 40 percent of the heat capacity per unit volume that water has. If an appropriate material can be found, latent-heat storage offers the advantage of avoiding unnecessarily high temperatures. For example, sodium sulfate (Glauber's salt) makes a transition at 32°C (90°F) between hydrated and nonhydrated crystals:

$$Na_2SO_4 \cdot 10H_2O \rightleftharpoons Na_2SO_4 + 10H_2O,$$

with a heat of reaction of 50 cal/g. As heat is added, the hydrated crystals will warm to 32°C. Then the material will remain at 32°C, absorbing heat but not changing temperature as the reaction proceeds to the right. Once the reaction has gone fully to the right (absorbing 50 cal/g), the temperature will rise as more heat is added. The reaction proceeds in reverse as heat is removed, delivering 50 cal/g heat of crystallization at 32°C. (In practice, the rate of crystallization is often too slow, and a supercooled solution tends to form.) With a heat of reaction of 50 cal/g and a density of 1.5 g/cm³, the latent heat storage capacity of Glauber's salt equals the sensible heat storage caused by a 75°C temperature rise of an equal volume of water. By keeping the heat storage temperature only slightly above the temperature needed for heat use (e.g., 68°F for home heating), one reduces heat losses from storage and at the same time increases the efficiency of the solar energy collectors.

A solar home-heating system is shown schematically in Fig. 5.8. Note that there is an auxiliary heat source—a conventional oil or gas furnace. It is not economical to install sufficient solar heating capacity to carry a house through several consecutive cold, cloudy days. In the several solar-heated houses described by Daniels,[3] the auxiliary furnace supplied from 5 to 75 percent of the annual heating requirements. Typical collector areas were about half the area

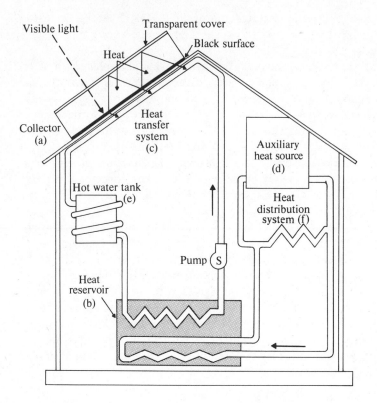

Fig. 5.8 Schematic view of a solar home-heating system showing (a) collector, (b) heat reservoir, (c) system to transfer energy from collector to reservoir, (d) auxiliary heat source, (e) hot water heater, and (f) heat distribution system. (From *Solar Energy Research,* Staff Report of the Committee on Science and Astronautics, U.S. House of Representatives, 92nd Congress, 2nd Session, December 1972. Washington, D.C.: U.S. Government Printing Office, 1973, pp. 78–79.)

of the house (ground floor area), i.e., in the neighborhood of 1000 sq ft. All were flat-plate collectors and operated at temperatures between 50°C and 60°C.

The few dozen existing solar-heated buildings demonstrate that solar home heating is feasible today using existing, conventional technology. It is difficult to evaluate its economic competitiveness from these cases, however, because all were custom built with accompanying high costs. Mass production of components, collectors in particular, is essential for competitiveness. Estimates are that mass-produced flat-plate collectors might sell for two to four dollars per sq ft. At these prices, solar heating is competitive with electric heating over a major portion of the United States. It is not competitive with oil or gas heating at oil and gas prices of a few years ago. A change in the relative prices of flat-plate collectors and oil/gas by a factor of two would make solar energy economically competitive in many areas of the country.

The task of generating electricity with heat from solar energy is more exacting than home heating. In the former case, the Carnot efficiency limit applies and high temperatures are necessary to obtain reasonable efficiency. It appears that focussing collectors are required. Meinel and Meinel [4] describe their concept of a solar-thermal-electric power plant in the 1000-MW$_e$ range. It involves focussing to obtain a factor of 10 concentration of the incident solar flux and selective absorption collectors ($a_V/\epsilon_{IR} \approx 10$). Collector temperatures of 500°C and collection efficiencies of 90 percent are envisioned. Energy storage would be by latent heat storage provided by heat of fusion of a salt eutectic (320,000 metric tons are required per day of reserve energy for a 1000-MW$_e$ power plant). Heat would be utilized in a conventional steam cycle at 40 percent efficiency. The 1000-MW$_e$ plant would occupy an area of 5 sq mi. My impression is that the "old timers" in the solar energy field view the Meinels' proposal as unduly optimistic, namely, that when all the details are accounted for, the overall efficiency will be considerably lower than indicated (see, for example, Hottel and Howard, 1972). Cost estimates are uncertain because neither the price nor the useful life of large quantities of selective absorption collectors are well established. The Meinels believe electricity at 5 mils/kWh might be achievable.

An alternate approach to electric power generation with solar energy is the use of *photovoltaic cells* (solar cells), which are semiconductor devices that convert light energy directly to electrical energy without passing through a thermal stage. Briefly, a photovoltaic cell consists of a p–n junction that can be exposed to light. Incident photons are absorbed, making electron-hole pairs in both the p and n type material. The energy levels of a p–n junction are such that the electrons created in the p-type material migrate across the junction to the n-type region, and the positive holes created in the n-type material migrate across the junction to the p-type region. In this way, an electric voltage (~0.5 volt) and an electric current are established. (The preceding explanation is assuredly incomprehensible to any not familiar with the rudiments of semiconductor physics.)

Silicon solar cells convert solar energy to electrical energy with an efficiency as high as 14 percent, while cadmium sulfide cells have a 4–5 percent efficiency. Silicon-solar-cell power plants for satellites cost a few hundred dollars per watt. This price must come down two to three orders of magnitude before solar cells can be of any competitive significance for terrestrial power. Cadmium sulfide cells are considerably cheaper than silicon cells, though still too expensive for terrestrial power. However, the history of semiconductor-device costs has been one of rapid and large price decreases, and an ultimate price of ten cents per watt has been envisioned.

Should solar cells become economical, three configurations for their use might be considered: (1) individual home units, taking advantage of the "free delivery" of solar energy, (2) large, central-station power generation, taking advantage of economies of scale, and (3) a satellite-based central power station, beaming power to earth in the form of microwaves. This last scheme has the advantage of continuously available, high-intensity solar radiation (no atmosphere, no clouds, no night), but there are so many other disadvantages that its success in competition with the other schemes remains improbable.

An important thing to realize about solar energy is that its utilization is technologically feasible today and has been for many years. The question is not whether it is possible, but whether it can be made *economical*. For specific uses (principally hot-water heating) in favorable locations (where there is good sunlight and high fossil-fuel costs), solar energy is economical today and is being used.

5.5   Fuel Cells

The conversion of fossil-fuel energy to electrical energy by means of an intermediate thermal energy stage has an overall efficiency of only 40 percent, while the theoretical limit imposed by the fossil fuel itself is quite close to 100 percent (see Section 5.1). In order to approach this limit with the thermal energy stage, it would be necessary to burn the fuel (and transfer the heat) at very high temperatures, a process the materials of present-day boilers cannot tolerate. This fact and others have given rise to considerable interest in *fuel cells,* devices which directly convert chemical energy to electrical energy without passing through a thermal energy stage.

Fuel cells operate on the same principle as batteries. They consist of two electrodes with intervening electrolyte. Externally supplied fuel is oxidized at one electrode, yielding electrons; externally suppied oxygen (e.g., from air) is reduced at the other electrode, consuming electrons. Since the chemical reactions at the electrodes are energetically favored, they will proceed until a considerable excess of electrons exists at one electrode and a considerable deficiency exists at the other, giving rise to a voltage difference between the electrodes. This voltage difference can cause current to flow in an external circuit, i.e., the system can deliver electrical energy. The conceptual difference between the fuel cell and the battery is that in the fuel cell, the chemical reactants are continuously supplied from outside the cell and the reaction products are continuously removed.

Fuel cells operating on hydrogen and oxygen have been used as power sources in the space program. Pratt and Whitney is developing fuel cells for terrestrial uses that use natural gas as fuel. They estimate that an economically competitive fuel cell will be available within three years. But the efficiency of these cells is only 40 percent. Westinghouse is investigating the possibility of fuel cells that operate at elevated temperatures ($\sim 1000°C$) and might have efficiencies greater than 60 percent, but they are experiencing difficulties with materials at the high temperatures.

It thus appears that low-temperature fuel cells are no more efficient than fossil-fuel steam-electric power plants. High-temperature fuel cells run into materials problems similar to those for thermal-electric generation. With improved materials, combined gas turbine-steam turbine thermal-electric plants are projected to approach 60 percent efficiency,[5] which is about as good as that projected for fuel cells. Since costs are likely to be higher and efficiencies no better than that for thermal-electric plants, fuel cells do not appear promising for central-station, baseline production of electricity. However, fuel cells should emit far less air pollutants than combustion-based power sources, and their waste heat can be readily released to the atmosphere. (Steam-electric plants usually give most of their waste heat to a river or lake.) Therefore, air-pollution and thermal-pollution considerations speak in favor of fuel cells.

The efficiency of large thermal-electric plants falls considerably if they are not operated at full capacity. Also, as plant size falls much below 50 MW_e, efficiency drops. For these reasons, fuel cells may have applications for peaking power at central-station facilities and for generation of small power levels (<1 MW_e) at remote locations where transmission costs and losses weigh against central-station use. Fuel cells have the advantage of being portable and conceivably could power electric automobiles.

5.6   Heat Pumps

A heat pump is a device which moves heat from a colder region to a warmer region if work is done on it. Our common experience with heat pumps is with refrigerators and air conditioners. There the interest is in removing heat from the region to be cooled (the inside of the refrigerator enclosure or the inside of the building). One can also use heat pumps to heat buildings by pumping heat from the outside air or some other convenient body into the building. It is that application that concerns us here.

A typical heat pump works just like a household refrigerator. Some working medium (e.g., freon) is carried around a cycle which consists of compression and liquefaction, accompanied by a rise in temperature and output of heat, followed by an expansion and evaporation, accompanied by a drop in temperature and absorption of heat.

Suppose by an expenditure of work W on a heat pump, Q_1 units of heat are removed from a body at a temperature T_1, and Q_2 units of heat are delivered to a body at a higher temperature T_2. Conservation of energy implies that $Q_2 = Q_1 + W$. We define the refrigeration efficiency η of the heat pump as

$$\eta = \frac{Q_1}{W}, \tag{5.33}$$

and the heating efficiency ϵ as

$$\epsilon = \frac{Q_2}{W}. \tag{5.34}$$

Note that $\epsilon = 1 + \eta$. By Carnot-theory considerations,

$$\eta \lesseqgtr \frac{T_1}{T_2 - T_1}, \tag{5.35}$$

$$\epsilon \lesseqgtr \frac{T_2}{T_2 - T_1}. \tag{5.36}$$

The larger the temperature difference through which the heat is to be pumped, the smaller the maximum efficiency. Because the heat Q_2 available for heating is the sum of the work done W and the heat removed from the cooler body Q_1, the heating efficiency ϵ is always greater than one.

Let us compare the overall fuel efficiency of three home heating schemes, (1) fossil-fuel electric generation, with electricity used to drive a heat pump, (2) fossil-fuel electric generation, with electricity used for resistance heating, and (3) home gas or oil furnace. Fossil-steam-electric generation has an efficiency of 40 percent; home furnaces are typically 75 percent efficient, with the remainder going up the chimney as heat and incomplete combustion. Resistance heating is 100 percent efficient in converting electrical energy to useful heat. The overall fuel efficiencies are therefore (1) heat pump—$0.40 \times \epsilon$, (2) resistive heating—0.40, and (3) furnace—0.75. The heat pump is always more efficient than resistance electrical heating, and is more efficient than the gas or oil furnace if ϵ exceeds two.

The Carnot limit allows ϵ's considerably in excess of two. For example, consider pumping heat into a house at 70°F (21°C or 294°K) from a body at 32°F (0°C or 273°K). Then

$$\epsilon_{\text{carnot}} = \frac{294}{21} = 14. \tag{5.37}$$

One would not expect to come very close to this efficiency in practice. The heat-transfer surfaces must be significantly above 70°F and below 32°F in order to obtain a reasonable transfer of heat per unit area. Also, the heat pump proper contains irreversible stages, and hence inefficiencies. Hirst and Moyers[6] find that heat-pump efficiencies (averaged over U.S. climatic conditions) are about two. One would hope that this figure could be improved.

A system for home heating with a heat pump can be installed in conjunction with a central air-conditioning system, since a large part of the equipment (e.g., the basic pump and the air-handling system) is common to both. Heat pumps for heating are most advantageous in warmer regions where the temperature differential in the winter is smaller (allowing larger values of η) and where a large-capacity heat pump is required for air conditioning anyway. Under these conditions, the installation costs attributable to heating are small. The number of households in the United States heated by heat pumps is in the hundreds of thousands.

5.7   Conversion of Solid Fuels to Fluid Fuels

Petroleum and natural gas are today's preferred fossil fuels for two reasons. They are easy to handle and they either contain fewer polluting impurities or can more readily have them removed. With reserves of these fuels dwindling, there is increasing interest in converting coal, oil shale, and tar sands to liquid or gaseous fuels. The principal components of all fossil fuels are hydrogen and carbon. As one progresses from solid to liquid to gaseous fuels, the hydrogen-to-carbon ratio rises: coal (~0.75), oil shale (1.6), tar bitumen (1.5), crude oil (~1.7), and methane (4.0). The basic problem in solid-to-fluid fuel conversion, then, is to increase the hydrogen-to-carbon ratio. For coal a large increase is required; for oil shale and tar bitumen only small increases are needed.

In most schemes for producing gaseous fuel, the supply of hydrogen is water vapor. Some of the relevant chemical reactions are:

$$C + H_2O \rightleftharpoons CO + H_2 \qquad -32,457 \text{ kcal/kg mole} \qquad (5.38)$$

$$CO + H_2O \rightleftharpoons CO_2 + H_2 \qquad +7,838 \text{ kcal/kg mole} \qquad (5.39)$$

$$C + 2H_2 \rightleftharpoons CH_4 \qquad +21,854 \text{ kcal/kg mole} \qquad (5.40)$$

The energetics of other reactions can be obtained by appropriate combinations of these three. For example, from Eqs. 5.38 and 5.39 we obtain

$$CO_2 + C \rightleftharpoons 2CO \qquad -40,295 \text{ kcal/kg mole.} \qquad (5.41)$$

A desirable overall reaction is

$$2C + 2H_2O \rightleftharpoons CH_4 + CO_2 \qquad -2,765 \text{ kcal/kg mole.} \qquad (5.42)$$

With a small addition of energy, carbon is converted to methane. This energy can be supplied by burning some of the carbon, i.e.,

$$C + \tfrac{1}{2}O_2 \rightarrow CO \qquad +26,637 \text{ kcal/kg mole.} \qquad (5.43)$$

The preceding suggests that one mix the solid fossil fuel (carbon) with steam and a small amount of oxygen (to supply heat energy via reaction 5.43), and arrange the temperature and pressure so that at equilibrium the various reaction products will be in desirable proportions. The reaction products in question are CH_4, CO_2, CO, and H_2. For most purposes, it is desired that the reaction

$$CH_4 + CO_2 \rightleftharpoons 2CO + 2H_2 \qquad -62,149 \text{ kcal/kg mole} \qquad (5.44)$$

be far to the left. That is, methane is the desired fuel. (Carbon monoxide and hydrogen gas can be used as fuels, but they have less energy content per unit volume, and therefore increased transportation costs per unit energy. Carbon monoxide is poisonous, arguing against its use in residential applications. For industrial applications with on-site or nearby gas production, $CO + H_2$ constitutes a satisfactory fuel.) At low temperatures, reaction 5.44 is far to the left and the net result is reaction 5.42, as desired. Unfortunately, at low temperatures the rate of chemical reactions (i.e., the rate of approach to equilibrium) is too slow to be useful. As temperatures are increased above 250°C, reaction 5.44 shifts to the right, and at temperatures at which the reactions proceed at a usable rate (~1000°C), the net result is reaction 5.38. The processes currently being developed thus produce $CO + H_2$ in a high-temperature stage, and then convert these gases to methane in a later, low-temperature stage, using catalysts to obtain usable reaction rates.

The general features of a scheme for producing methane from coal are shown in Fig. 5.9. In the first stage, coal is heated to drive off its volatile components. The nonvolatile residue (carbon plus ash, called *char*) is transported to a second stage, where it is heated further and reacts with steam to form gas, largely $CO + H_2$. The gas streams from both stages require purification. Natural

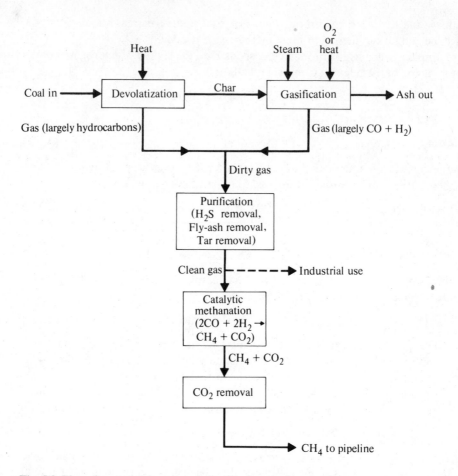

Fig. 5.9 Flow diagram of the process of producing gas from coal.

gas is quite free of sulfur and ash, but the manufactured gas, at this point in the process, is not. (Purification is much easier for this gas than it is for combustion-product gases from coal combustion (e.g., stack gas cleanup in a power plant), in part because the volume of gas per unit energy is smaller, and in part because sulfur is in the form H_2S rather than SO_2.) The purified gas goes to a catalytic methanation stage, and the output from this stage goes to a carbon dioxide removal stage. The final output is methane, equivalent to natural gas. If the gas is intended for industrial use or for generation of electricity, the last two stages can be eliminated.

In actual practice, there are several variations on the theme just presented, as well as a host of engineering difficulties to be overcome. As of now there are pilot plant projects for methane from coal, but no full-scale commercial plants. Significant production is not expected before 1985.

Schemes for converting coal to a liquid fuel entail dissolving the coal in a solvent, adding hydrogen, and adjusting temperature, pressure, and catalysts so that hydrogenation occurs. Sulfur (in the form H_2S) and ash are removed, and the solvent is separated for reuse. The "coal," increased in hydrogen-to-carbon ratio and now a heavy liquid, is usable as a boiler fuel or can be refined to make a variety of petroleum products. During the present research and development stage, hydrogen can be supplied from natural gas. For commercial applications, hydrogen would be made by reacting steam with coal, as described earlier.

Two other approaches should be mentioned. (1) By heating coal in the presence of its volatile products, a process called *pyrolysis,* one can obtain gas, oil, and char. The relative heat contents are 17 percent, 19 percent, and 56 percent, respectively, with 8 percent lost in the overall process. While the overall efficiency is high, 56 percent of the fuel remains in solid form. (2) In the *Fischer-Tropsch process,* coal is converted to gas, as described earlier, and the gas is used to synthesize a range of liquid hydrocarbon products. While this is the only process operating on a commercial scale (in the Republic of South Africa), costs are very high by U.S. standards. The other processes are at the pilot-plant stage and are less advanced than processes for gas from coal.

Tar sands and oil shale were discussed in Section 4.4. After separation from sand, tar bitumen is still too viscous to pipe any appreciable distance. However, bitumen requires only a small increase in hydrogen-to-carbon ratio, and this can be readily accomplished by standard petroleum-industry techniques. In the case of oil shale, kerogen is separated from the associated rock by heating. Heating decomposes the kerogen and volatile components are driven off. Heat is supplied by burning the small amount of nonvolatile residue (carbon). Since the hydrogen-to-carbon ratio for kerogen is quite high, its conversion to a petroleum-like material is relatively easy. As with tar bitumen, a small amount of "upgrading" is required to reduce viscosity. Environmental problems associated with shale-oil production are discussed in Section 6.23.

Questions and Problems

5.1
Convert the fission energy release of 1 MeV per nucleon to a figure in joules per kg of ^{235}U; to a figure in kWh per kg of ^{235}U. Similarly, convert the fusion energy releases of Eqs. 5.23 and 5.24 to kWh per kg of deuterium, and the fusion energy release of Eq. 5.25 to kWh per kg of lithium-6.

5.2
Consider a chain reaction with an effective reproduction constant $k_e = 1.1$.

a) If there are ten fissions in the first generation, how many will there be in the second generation? the 3rd? the 10th? the nth?

b) How many generations are required for the number of fissions per generation to increase by a factor of 1000? 10^6? 10^{12}?

5.3
Consider a neutron of mass m and initial kinetic energy E_i scattering elastically from a nucleus of mass M initially at rest. (By elastic scattering, one means that only kinetic

energy is involved; none of the initial kinetic energy is transformed into energy of nuclear excitation.)

a) For a head-on collision with the neutron scattering backwards (through 180°), what fraction of its initial kinetic energy does the neutron lose? Plot fractional energy loss vs. m/M.

b) More generally, for scattering through an angle θ, what fraction of the initial energy is transferred from the neutron to the nucleus? Note the dependence on m/M. (*Hint:* Use conservation of momentum and kinetic energy.)

5.4 *

a) Derive a differential equation for the time dependence of the power generated by a reactor (the fissions per second) $P(t)$ in terms of the reproduction constant k_e (assumed constant and slightly different from 1), the time between generations for prompt neutrons τ, the fraction of neutrons that are delayed f, and the mean delay time T. (Assume $k_e \approx 1$, $f \ll 1$, and $\tau \ll T$. Assume that, aside from their delay in emission by a time T, the delayed neutrons behave the same as prompt neutrons, i.e., the same fraction of them give rise to fissions.) Solve the differential equation and find $P(t)$. Divide your derivation into three cases:

1. Assume that all neutrons are prompt (i.e., set $f = 0$).
2. Allow for the existence of delayed neutrons, and assume that the power level changes by a small fraction of itself during the time interval T (i.e., $T dP/dt \ll P$).
3. Consider the case where prompt neutrons alone are sufficient to cause a divergent chain reaction, making the inequality of (2) above invalid, though $\tau dP/dt \ll P$ is still true.

b) For cases (2) and (3) above, identify a characteristic time, i.e., that time required for P to change by a factor e. What condition (in terms of k_e, f, τ, and T) must be met for the inequality in case (2) to hold? What condition must be met for a divergent chain reaction to proceed via prompt neutrons alone?

5.5 * ‡

Resonance capture cross sections for neutrons on nuclei at rest are well described by the expression

$$\sigma(E) = \frac{\sigma_0 \left(\dfrac{\Gamma_0}{2}\right)^2}{(E - E_0)^2 + \left(\dfrac{\Gamma_0}{2}\right)^2},$$

where E is the energy of the incident neutron, and σ_0, Γ_0, and E_0 are constants which describe the particular resonance. E_0 specifies the location of the resonance, Γ_0 its width, and σ_0 the cross section at the peak of the resonance. For uranium-238, the following resonances are important:

E_0 (eV)	Γ_0 (meV)	σ_0 (barns)
6.68	26.5	22,000
21.0	34.0	25,000
36.8	59.0	18,000
66.3	43.0	10,000
103.0	90.0	4,200
117.0	34.0	5,800

How is the expression for the cross section modified if the target nuclei are not at rest, but rather in random thermal motion at a temperature T?

5.6 *
Consider a reactor adjusted to operate at a steady power level P_0 at a temperature T_0. If the power produced P differs from the power withdrawn P_0, the temperature of the reactor will change:

$$\frac{dT}{dt} = \alpha(P - P_0),$$

where α is a positive constant. (Justify this equation.) If the temperature differs from T_0, the effective reproduction constant k_e will change from unity:

$$k_e = 1 - \beta(T - T_0),$$

where β is a positive constant. (Justify this equation.)

a) Obtain a second-order differential equation satisfied by P, and obtain an approximate solution. Comment on stability.

b) Now assume the power withdrawn is not a constant, but equals $P_0 T / T_0$, as would be a good approximation if the flow rate of water through the reactor was kept constant. How do the equations change?

5.7
Derive Equation 5.28. For a start, consider one layer of glass only ($n = 1$) and ignore the ambient radiation ($T_0 = 0$), but allow for a useful energy flux q_0 being drawn from the collector. Next, allow for the fact that the collector receives IR radiation from nearby objects at temperature T_0, even if the solar flux S is absent. Finally, generalize to the case of n layers of glass.

5.8
Derive Eq. 5.31. For a start, ignore the ambient radiation ($T_0 = 0$) and draw no useful energy from the collector ($q_0 = 0$), i.e., derive Eq. 5.30. Next, let useful energy q_0 to be drawn. Finally, allow for IR radiation received from nearby objects at an ambient temperature T_0.

5.9
For an idealized flat-plate collector with one layer of glass (Eq. 5.29), plot collection efficiency η vs. collector temperature T. Repeat for a collector with four layers of glass. For an idealized flat-plate collector making use of a selectively absorbing surface (Eq. 5.32), plot η vs. T for $a_V = 1.0$, $\epsilon_{IR} = 0.1$; for a_V 1.0, $\epsilon_{IR} = 0.05$.

5.10
The overall efficiency of a solar energy system is the product of the collection efficiency η and the efficiency of utilization. The efficiency of utilization depends on the temperature of the collector and on the intended use for the energy. Consider two uses, (a) heating, and (b) doing work.

a) If the intended use of the collected energy is to maintain a home at a temperature $T_1 = T_0 + 30°C$, then the efficiency of utilization is zero for $T < T_1$ and can in principle be 100 percent for $T > T_1$. For the four collector configurations described in Problem 5.9, obtain the maximum overall efficiency for home heating.

b) If the intended use of the collected energy is to do work (generate electricity, drive an engine, etc.), then Carnot theory gives the upper limit on the efficiency of utilization, with the Carnot engine running between the collector temperature T and the ambient temperature T_0, assumed $\sim 270°K$. For the four collector configurations described in Problem 5.9, obtain the maximum overall efficiency for doing work. (*Warning:* These figures are upper limits for efficiencies, based on idealized collectors. Practical efficiencies will be significantly lower.)

5.11
In heating a house with solar energy, it is necessary to store sufficient energy to heat the house from sunset to sunrise, a period of about 16 hours in the winter in mid-latitudes. A typical well-insulated house loses 36,000 Btu/hr if the outside temperature is 70°F below the inside temperature (see Table 8.2). If water is used as the energy storage medium, how much water is needed if a 60°F rise in water temperature is allowed? a 120°F rise? What are the pros and cons of utilizing a large temperature rise as opposed to a small rise?

5.12 * ‡
Consider a perfect, selective-surface, flat-plate solar collector, with $\epsilon = a = 0$ for frequencies $\nu < \nu_0$, and with $\epsilon = a = 1$ for frequencies $\nu > \nu_0$. Assume there are no losses from the collector due to conduction or convection, and that the back side of the collector is perfectly reflecting and so does not radiate. Assume the collector is placed in sunlight (energy flux $= 1$ kW/m², and frequency distribution appropriate for a black body at 5800°K).

a) For several values of ν_0, determine the collection efficiency (useful power drawn/incident solar flux) as a function of collector temperature. Determine the collector temperature when no useful power is being drawn, and plot *vs.* ν_0.

b) Assume the energy drawn from the collector drives a Carnot engine whose low temperature reservoir is at 300°K. Obtain the maximum theoretical efficiency for converting diffuse sunlight into work. (*Hint:* This problem is most easily solved numerically with a computer.)

5.13 *
Determine the basic parameters of a solar home heating system, specifically, the collecting area (use flat-plate collectors with one layer of glass) and the volume of the heat storage reservoir (use water). Assume 700 Btu/ft²-day as the solar flux on a typical winter day. (The U.S. year-round average is 1400 Btu/ft²-day.) Assume 18,000 Btu/hr as the heat loss from the house on a typical winter day. (This value is appropriate for a well-insulated house heated 35°F above the outdoor temperature.)

a) First, determine the collecting area and storage capacity appropriate for a typical winter day (24-hour period). Describe how the system would function throughout such a day, giving the temperature of the storage tank throughout the 24-hour period, assumed collector temperature and efficiency, and so on.

b) Next, consider the modifications necessary due to the fact that all days are not typical, some being considerably colder than average and some having considerably lower solar fluxes than average. Suppose the system was to handle three consecutive cold, heavily clouded days, preceded by only one typical day. What increases in storage capacity and collector area would be required?

Notes

1
Appendix B contains a brief refresher course on nuclear physics, including a discussion of fission.

2
Clarence Zener, "Solar Sea Power." *Physics Today* **26,** No. 1, p. 48.

3
Farrington Daniels, *Direct Use of the Sun's Energy.* New Haven: Yale University Press, 1964, p. 154.

4
A. B. Meinel and M. P. Meinel, "Physics Looks at Solar Energy." *Physics Today* **25,** No. 2, p. 44.

5
H. C. Hottel and J. B. Howard, *New Energy Technology—Some Facts and Assessments.* Cambridge, Mass.: MIT Press, 1972, p. 281.

6
E. Hirst and J. C. Moyers, "Improving Efficiency of Energy Use: Transportation and Space Heating and Cooling." In *Energy Research and Development,* Hearings before the Subcommittee on Science, Research, and Development of the Committee on Science and Astronautics, House of Representatives, 92nd Congress. Washington, D.C.: U.S. Government Printing Office, 1972, p. 513.

Related Reading

H. C. Hottel and J. B. Howard
New Energy Technology—Some Facts and Assessments. Cambridge, Mass.: MIT Press, 1972. Treats all topics of Chapter 5 except fusion and heat pumps.

A. Hammond, W. Metz, and T. Maugh
Energy and the Future. Washington, D.C.: AAAS, 1973. Twenty-two articles based on a series in *Science.* Most deal with energy technology.

Timothy J. Healy
Energy, Electric Power, and Man. San Francisco: Boyd and Fraser, 1974.

Walter Hossli
"Steam Turbines," *Scientific American* **220,** No. 4.

F. L. Cullen and W. O. Harms
"Energy from Breeder Reactors," *Physics Today* **25,** No. 5.

G. T. Seaborg and J. L. Bloom
"Fast Breeder Reactors," *Scientific American* **223,** No. 5.

Richard F. Post
"Prospects for Fusion Power," *Physics Today* **26,** No. 4.

J. Nuckolls, J. Emmett, and L. Wood
"Laser-Induced Thermonuclear Fusion," *Physics Today* **26,** No. 8.

J. Emmett, J. Nuckolls, and L. Wood
"Fusion Power by Laser Implosion," *Scientific American* **230,** No. 6.

Farrington Daniels
Direct Use of the Sun's Energy, New Haven: Yale University Press, 1974.

Solar Energy Research, Staff Report of the Committee on Science and Astronautics, U.S. House of Representatives, 92nd Congress, 2nd session. Washington, D.C.: U.S. Government Printing Office, 1973.

"An Assessment of Solar Energy as a National Energy Resource," NSF/NASA Solar Energy Panel, December 1972.

Harry Perry
"The Gasification of Coal," *Scientific American* **230,** No. 3.

Derek P. Gregory
"The Hydrogen Economy," *Scientific American* **228,** No. 1.

6 Energy-Related Environmental Problems

> We are remodeling the Alhambra with a steam-shovel, and we are proud of our yardage. We shall hardly relinquish the shovel, which after all has many good points, but we are in need of gentler and more objective criteria for its successful use.
>
> A. Leopold, *A Sand County Almanac**

This chapter is a long one, and is therefore divided into five parts. Part A discusses environmental problems generally and uses as illustrations problems not necessarily related to energy. At certain points, this treatment will appear "soft," subjective, philosophical, because there are aspects of environmental problems which cannot yet be handled any other way. But these aspects are, in my opinion, of basic importance—better to treat them in a "mushy" way than to omit them entirely.

A large fraction of our current and anticipated environmental problems are quite directly related to energy. *Thermal pollution,* the addition of heat to the earth's environment as a result of human activity, is completely due to energy utilization. While much of present man-made *radioactivity* is due to testing of nuclear weapons, projections into the future show energy utilization as the dominant source. About 75 percent of man-made *air pollution* is caused by energy utilization. These three topics are treated in Parts B, C, and D. Because each is an appropriate subject for a complete book, our treatment necessarily will be selective and simplified. Several other energy-related environmental problems are lumped together, comprising Part E.

---

* From *A Sand County Almanac, With Other Essays on Conservation from Round River* by Aldo Leopold. New York: Oxford University Press, 1966, p. 241. By permission.

PART A / GENERAL CONSIDERATIONS

6.1   What Are "Environmental Problems"?

A list of some of the items of concern to environmentalists is given in Table 6.1. The list is diverse, ranging from urban decay to wilderness preservation. By

Table 6.1   Partial list of items of concern to environmentalists.[a]

Water pollution	Visual pollution
Air pollution	Wilderness preservation
Resource depletion	Wildlife preservation
Radioactivity	Urban decay
Solid-waste disposal	Suburban sprawl
Thermal pollution	Agricultural malpractices
Pesticides	Feed lots—"sewage" disposal
Land use	Fertilizer runoff
Energy utilization	Soil misuse—erosion
Transportation—mass transit	Bad lumbering—clear cutting
Noise pollution	Oil spills

[a] (Some items on this list overlap. For example, fertilizer runoff is an example of water pollution; land use includes aspects of urban decay, suburban sprawl, and wilderness preservation. Further, many items on the list are related to one another, and so individual items cannot be dealt with in isolation.)

analyzing the items on this list, four types of environmental damage can be identified:

a) Threats to human health and safety
b) Damage to economic resources and to material well-being
c) Reduction in "enjoyment of life" of a psychological/aesthetic character
d) Damage to nonhuman environments—to nature.

Examples of the first category, direct public health problems, include air pollution which can cause lung disease, radioactivity which can cause cancer, water pollution which can cause typhoid fever, and noise which can cause hearing loss.

Damage to economic resources can be direct. Air pollution can damage crops and timber, paint on buildings, and stone on buildings. Water pollution increases the cost of water purification for home and industry. Nitrate fertilizer runoff makes water supplies unfit to drink.

When the economic resource is part of a natural ecosystem, the damage can also be indirect. In this case, some other part of the ecosystem is damaged, which, through the interdependencies of the various parts, leads to damage to the economic resource. Many commercial fish depend on a food chain, some of whose links spent part of their life cycle in coastal marshes and estuaries. Filled marshes and polluted estuaries thus damage commercial ocean fisheries. Similarly, freshwater pollution can damage freshwater fisheries indirectly by damaging a link in a food chain. Insecticide use can eliminate the natural predators of some insect pest. The pest then increases rapidly in number and may damage timber resources or crops. Methyl mercury in aquatic ecosystems becomes concentrated as it moves up the food chain, and may reach such concentrations in the commercial fish at the top of the food chain that they become unsafe for human consumption. Natural ecosystems are sufficiently complicated and poorly understood that damage of an indirect nature is often unanticipated. Its cause is recognized only after much damage is done and corrective action difficult or impossible.

Categories of reduction in "enjoyment of life" include (1) aesthetic damages such as billboards along an attractive road or electric transmission lines marring a landscape, (2) psychological discomforts such as noise or odors, and (3) diminished recreational opportunities. In this last category, upsetting natural ecosystems very frequently reduces outdoor recreational opportunities. Stream siltation caused by exploitive lumbering damages sports fishing; summer home developments destroy woodland, thus reducing hunting and hiking recreational values; ocean oil pollution messes up beaches, thus diminishing our enjoyment of them.

Examples of damage to nature include the following. DDT, which becomes more concentrated as it moves up the food chain (because of its long life, and solubility in fat but not water), is causing breeding failures in peregrine falcons and other raptorial birds, threatening their extinction. Predator control programs, which carry on poisoning and offer bounties, have very nearly eliminated wolves from the continental United States. Lumber harvesting in mature redwood forests has significantly reduced the acreage of such forests. Extensive filling and draining of swamps and marshlands has greatly reduced the extent of that type of ecosystem, with an accompanying reduction in certain species of plants and animals.

Environmental problems can be placed on a man-nature continuum, with "harm to humans only" at one end and "harm to nature only" at the other end (see Fig. 6.1). Concerning this continuum:

1. The *extent* of present damage in the United States becomes progressively more severe as one moves from the human-only end toward the nature-only end. For example, the health and general well-being of Americans has improved considerably over the past 100 years; the health and well-being of U.S. commercial forests has remained generally static; that of hunting and fishing grounds has deteriorated somewhat; while the health and well-being of wetlands, redwood forests, and grizzly bears has deteriorated markedly.

2. The *importance* of damage becomes more debatable (at least more debated!) as one moves from the human-only end toward the nature-only end.

3. It is argued by some that the nature-only end of the continuum does not exist, that man is dependent on nature, so that any damage to nature hurts man, often in unexpected ways. Stated differently, any example of type (d), it is claimed, will prove also to be an example of type (a), (b), or (c) on close inspection. The harm may involve loss of a benefit currently received by man, or the closing of an option for a future benefit. The usual arguments talk about food supply, air and water purification, and other physical needs (type (a) or (b)).

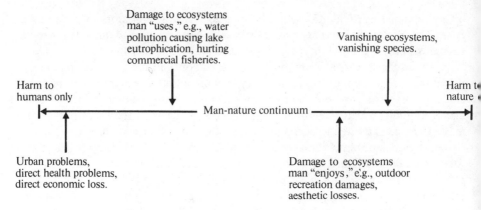

Fig. 6.1 Man-nature continuum of environmental problems.

(For example, whales eat very tiny plants and animals and convert them to a food supply that is usable by humans. Because this food chain is short, whales represent a good "machine" for harvesting food from the sea. Though past and current emphasis in whaling has been to obtain whale oil, in the future it could be to obtain food for human consumption. The impending extinction of whales would close this option.) I do not find these arguments entirely convincing. It is certainly true that many damages to nature cause man substantial material harm and limit his future opportunities for material well-being. However, I suspect that there are cases where there is major damage to some natural ecosystem or elimination of some species, but only minimal *material* loss to man either now or in the future. Examples might be the significant reduction in the acreage of climax redwood forests, or the near elimination of grizzly bears from the continental United States. In these cases, it may be said that there are significant aesthetic/psychological losses to man, i.e., type (c) damage. Further, even when type (a) or (b) damages do exist (e.g., extinction of whales), the type (c) damages may be more significant.

4. The approaches that different people take to an impending environmental damage which appears to affect only nature include the following. (1) If the damage cannot be demonstrated as affecting humans in a substantial material way, it can be ignored. (2) The damage should be avoided if there exists the possibility that it will affect man's physical well-being, either now or in the future, in ways we do not now fully understand. (3) The psychological/aesthetic loss to man is an important consideration and constitutes grounds for preventing the damage. (4) The nonhuman world has rights also, independent of any benefit to man, and therefore should be protected.

Many of the items listed in Table 6.1 appear in several places on the continuum of Fig. 6.1 and represent more than one type of harm. For example, pesticides can cause acute poisoning and death in accidents of application, chronic poisoning causing health problems from low-level, continuous ingestion of residues on foods, damage to livestock in accidents of application, damage to commercial fisheries from their presence in aquatic ecosystems, damage to sports fishing in the same way, and finally, damage to bird species of no apparent value to man.

Environmental problems are invariably due to quantitative misuse. The Indian boy in pre-Columbian times standing at the side of a stream and relieving himself into it did not constitute an environmental problem; the present-day city dumping its sewage into the same stream does constitute an environmental problem. The stream can assimilate wastes up to some level with no damage—beyond that level, wastes dumped into the stream become an environmental problem. The same situation applies to many other pollutants—natural processes can assimilate them up to some level, but beyond that level damage occurs and becomes more severe as the level is increased. Similar arguments apply to environmental problems other than pollution. For example, filling in the first 5 percent of the coastal wetlands along the Atlantic coast was not of much environmental concern, but filling in the last 5 percent will be an environmental catastrophe. Similar comments apply to lumbering climax redwood forests, damming free-flowing rivers, or destroying any other particular type of ecosystem.

Since environmental problems are quantitatively based, we must continually ask, "How much is too much?" But ecosystems are very complex and our knowledge about them rather meager. Since nature is a working system, it behooves us to be very careful in tampering with it. As a rule of thumb, we can use a "one-percent law"—when anything man is doing approaches 1 percent of what nature is doing (on a regional scale), think very carefully before proceeding.

Why pick 1 percent, rather than, say, 0.1 percent or 10 percent? From the individual cases I've looked into, a 0.1-percent human intervention rarely has significant effects, while a 10-percent one often does. Note that the rule doesn't say, "Stop at 1 percent," but rather that it says, "Think very carefully before proceeding." What about human activities for which there are no natural counterparts? (For example, all DDT found in natural ecosystems is of man-made origin, so the original use of DDT represented an infinite percentage increase over nature.) Does this mean that before one enters into any new technology involving impacts on ecosystems of a variety that have no natural counterparts, one should "think very carefully"? Damn right it does!!

6.2 Cost-Benefit Analysis

In any consideration of specific environmental problems, there are tradeoffs. In deciding to build a nuclear power plant rather than a fossil-fuel plant, one is trading air pollution for radioactivity. In legislating strict emission standards for automobile exhausts, one is trading the environmental costs of air pollution for the dollar costs of the abatement program. In foregoing lumbering a wilderness area, one is trading the economic value of the timber for the psychological/aesthetic/recreational value and natural value of wilderness.

The recommended approach in choosing between alternative courses of action is to perform a *cost-benefit analysis*. For each alternative, all the costs (dollar and otherwise) and all the benefits (economic and otherwise) are quantitatively evaluated. These costs and benefits are converted to some common denominator (usually dollars) and the different alternatives compared. That alternative with the largest net benefit (benefit minus cost) is chosen, thereby maximizing net benefits.

The above sounds simple in principle but difficult in practice. In fact, there are difficulties of principle as well as practical problems.

1. The costs often accrue to people different from those who receive the benefits. For example, the environmental costs of strip mining reside in Appalachia, while the benefits of cheap coal go mainly to the city dwellers of the Northeast. Often the costs and benefits accrue to different generations. For example, most acid mine drainage is caused by abandoned coal mines, some abandoned more than a century ago. Another more extreme example is the nuclear power plant. Its benefit, electricity, is provided over a 30 to 40-year period, the lifetime of the plant. But one of its major environmental costs, guaranteeing the safe storage of radioactive wastes from the plant, must be borne for a 1000-year period.

2. It is difficult to quantify some of the costs and benefits, particularly those of types (c) and (d) (see page 104). For example, how can one assign a dollar value to the aesthetic loss caused by an attractive roadside landscape being marred by electrical transmission lines? By estimating the reduction in the sales value of the property? This certainly underestimates the cost, because the aesthetic losses to most people (e.g., those driving by) would not be reflected in changed property value. It is tempting to give up on estimating costs or benefits not readily quantifiable and just ignore them. This temptation must be resisted, because ignoring an item is equivalent to assigning a quite definite value to it— namely, zero. It is equally unwise to insist that some feature must be protected "no matter what." That is equivalent to assigning a value of infinity to it. Any reasonable estimate of the aesthetic value of a landscape will be neither zero nor infinity.

Many people believe that costs and benefits of a psychological/aesthetic or "nature-only" variety are *in principle* unquantifiable. But a philosophical postulate of modern science says that if something cannot in principle be observed or measured, it does not exist, i.e., its existence or nonexistence is of no consequence. (If something has consequences, then these consequences can provide a means for obtaining a measurement.) Since costs and benefits of types (c) and (d) clearly are of consequence, the philosophical position of modern science would be that they are in principle quantifiable. An important unfinished (indeed practically unstarted) task is to find techniques for this quantification. Such techniques will probably draw as heavily on psychology as on ecology.

Though modern science postulates that psychological/aesthetic and nature-only costs and benefits can (in principle) be quantified, it does not guarantee that the quantified results can be directly compared with similarly quantified public-health or economic costs and benefits. Within the realm of science, these may well be incommeasurable. Comparing human lives, dollars, beauty, and species diversity probably cannot be done by scientific techniques alone.

3. The cost-benefit analysis must be performed on the basis of incomplete knowledge and information. Our general knowledge about ecosystems, about social phenomena, and about many other fields relevant to each possible course of action is in many instances very rudimentary, and data appropriate to the specific case at hand often quite meager. Since human actions are much more likely to damage an ecosystem than to improve it (unless the action is taken with the specific purpose of improving the ecosystem), this lack of knowledge tends to result in an *underestimate* of the environmental costs of human actions. Similarly, many people believe that we receive benefits from natural ecosystems of which we are unaware, and this lack of knowledge tends to result in an underestimate of the benefits of refraining from the human action. Stated more simply, when people set out to do something, they are usually much more aware of the good things their action will accomplish than of the bad things. The incompleteness of knowledge does not randomly influence the outcome of cost-benefit analyses, but rather biases the decision in favor of increased human intervention into natural processes.

4. A cost-benefit analysis is made in a changing world and must involve some assumptions about those changes. If the world develops differently from the assumptions, the analysis will be defective. For example, the conclusion that it is of maximum benefit to lumber a particular wilderness area may rest on the assumption that certain other areas will remain inviolate. If these other areas are subsequently developed, the analysis will have been incorrect. Similarly, the decision not to build an electric power plant may rest on the assumption that certain energy-demand-reducing steps (e.g., energy conservation measures and rate structure changes) will be taken at some future time. The correctness of the analysis depends on whether or not these steps are actually taken. Environmental decisions are interdependent with a myriad of other decisions, and dealing with them one at a time is an imperfect approach at best.

I have the following subjective impressions of analyses and arguments on environmental issues.

1. Costs and benefits of a public health nature (type (a)) are given too much weight, while costs and benefits of a psychological/aesthetic variety are given too little weight. (This statement refers to the United States, with its high standards of public health and material well-being.) Individual human decisions do not show this emphasis. A significant number of people engage in dangerous recreational activities (mountain climbing, sports car racing, skin diving), and the average individual conducts his life with a view more toward making it enjoyable than extending its length. The fraction of personal income going toward subsistence items is not major, and the amount going toward "quality-of-life" items is appreciable. The amount of time devoted to recreation is also appreciable. It appears that the individual is attempting to maximize his or her happiness per unit time rather than total years lived. Perhaps cost-benefit analyses should similarly give increased weight to recreational opportunities and "quality of life" and attempt to maximize happiness per individual-year, rather than stress lifetime per individual or material goods per individual-year.

The importance of benefits and costs of types (c) and (d) relative to those of types (a) and (b) is very eloquently discussed by Aldo Leopold in *A Sand County Almanac*. The following nicely states his point of view:

> There are some who can live without wild things, and some who cannot. These essays are the delights and dilemmas of one who cannot.
>
> Like winds and sunsets, wild things were taken for granted until progress began to do away with them. Now we face the question whether a still higher "standard of living" is worth its cost in things natural, wild, and free. For us of the minority, the opportunity to see geese is more important than television, and the chance to find a pasque-flower is a right as inalienable as free speech.
>
> These wild things, I admit, had little human value until mechanization assured us of a good breakfast, and until science disclosed the drama of where they come from and how they live. The whole conflict thus boils down to a question of degree. We of the minority see a law of diminishing returns in progress; our opponents do not.[1]

2. Many environmentalists feel obligated to argue their cases (e.g., for preservation of some ecosystem) on public health or economic grounds, when their personal convictions are really based on psychological/aesthetic or "rights-of-nature" considerations. Thus some of their arguments appear rather contrived.

3. The importance of *variety* in the total environment is not adequately appreciated. Different people like different things, and the same people like different things at different stages in their lives. Indeed, an individual likes many different things at the same time. A very diverse environment is required to provide opportunities for all these "likes": cities, rural living, urban nightlife, wilderness, amusement parks, wetlands, art galleries, concert halls, deserts, mountains, trails for skimobiles, trails for cross-country skiing, lakes for motorboating, lakes for canoeing, places where people can get together, and places where people can be alone. A high quality of life, a maximum happiness per man-year, thus requires diversity. Another argument for diversity is a generalization of the ecological finding that complex, diverse ecosystems are usually more stable than simple, uniform ecosystems. If something goes wrong, the diverse ecosystem can resort to more alternative patterns. The relevance of a plea for variety here is that the usual decision-making process tends to follow precedents, a tendency that leads to uniformity rather than variety. For example, one of the benefits invariably claimed for the construction of a dam on a river is the recreational value of the lake thereby created. For the first dam in a region with few lakes and many free-flowing rivers, this is a legitimate claim, because the variety of the region is increased. As subsequent dams are built, lakes become commonplace and stretches of free-flowing rivers rare. Building another dam then reduces the variety of the region; the claim of recreational benefits is no longer valid.

6.3   Approaches to Solutions

It may seem surprising at first sight that self-interest did not lead to early solution to many of our present-day environmental problems. Indeed, it is an article of faith of the free-enterprise system that the sum of all individual "marketplace" decisions, arrived at on the basis of self-interest, leads to an optimization of the collective interests of society. This article of faith can be "proven" (to some people's satisfaction) assuming the following conditions: (1) there is a market; (2) there is a large number of competing buyers and sellers; (3) buyers and sellers have full knowledge of their opportunities and the consequences of their choices, and (4) the buyer and seller in a given transaction receive all the benefits and bear all the cost. But in the case of most environmentally related decisions, condition (4) is not met. In the jargon of the economist, there is an "externality," an "external cost," namely, a cost borne by someone who is not a part of the transaction. Under this situation, the free-enterprise system does *not* optimize the collective interest.

The solution most appropriate within the free-enterprise framework is to "internalize the externalities," that is, to modify the rules of the game so that the costs and benefits accrue to the decision maker rather than to someone else. Consider, for example, the case of reclaiming land damaged by strip mining for coal. If the land is not reclaimed, an environmental cost is borne principally by the people living in the general area of the mine. If the land is reclaimed at government expense, a dollar cost is borne by the taxpayers. Both cases are examples of externalities. This externality can be internalized by requiring any company engaged in strip mining for coal to return the land surface to the condition it had prior to mining. In reclaiming the land, the company incurs a dollar cost, which it takes into account in any of its transactions, i.e., the sales price of coal rises. When externalities exist, the price of coal is lower than it "ought" to be, and consumers will buy more than is optimum for the general welfare. The tax cost for land reclamation that an individual consumer pays is based principally on the decisions of *other* consumers to buy coal and very little on his own decision. The result is that the average consumer will buy more coal and pay more money out for taxes plus coal purchases than is optimum. If the land reclamation costs are included in the price of coal, the average consumer will buy less coal and spend less money, thus achieving optimum conditions for society collectively.

Internalization of externalities is a fine approach, but only if it can be achieved. However, it is usually difficult, if not impossible, to accomplish. *External costs are an intrinsic, essential feature of most, if not all, environmental problems.* In some cases, they can be internalized to some extent by laws and regulations. In most cases they cannot. (In the strip mining example, one of the easier cases, the mining company would be motivated to do only the absolute minimum reclamation required by law. In other words, it would attempt to externalize costs if it could.) It is a useful exercise to consider various environmental costs and try to develop schemes for internalizing them. The difficulty in internalizing environmental costs and benefits is closely related to one of the difficulties in cost-benefit analysis discussed in Section 6.2, namely, that costs and benefits accrue to different people.

There are problems with the other conditions necessary for the proper functioning of the free-enterprise system. Knowledge is usually incomplete and competition in some cases is limited.

With self-interest generally failing as a motivating force for environmental improvement, other devices must be used. A principal one is government control and regulation, in the form of air-pollution emission standards, water-pollution emission standards, land-use plans, etc. Certainly controls and regulations will play a major role in encouraging environmental improvement for the foreseeable future. But this approach has the defect that it is no better than the particular set of regulations being used. Individuals, corporations, and municipalities are motivated to satisfy the letter of the rules rather than their intent. Environmental ill-effects are exported to that state with the most lenient regulations. Actions are modified to conform to the regulations and minimize dollar costs, with the probable consequence of large environmental costs in areas not covered by the regulations. It is not possible to construct a set of general rules, that is sufficiently complete as to be optimal for each situation where these rules are applied.

An alternative approach, applicable for large projects, is project review by some appropriate regulatory board, based on a cost-benefit analysis of the proposed project and any alternative courses of action (i.e., an environmental impact statements; see Section 6.4). This approach sounds fine where it is applicable (i.e., big projects). It is too early to draw any conclusions about how it works in practice. (In the past, regulatory agencies have fallen under the influence of the group they were intended to regulate. To prevent this from happening with environmental matters, an attempt is being made to have more citizen participation in the review process.)

Many believe that external coercion of any sort is not adequate to the task of supplying motivation for environmental improvement. They argue that what is required is education, a change in values, and a change in morals and ethics. Individuals must want to "be kind to nature" just as they want to "love thy neighbor." The theme of the need for a new ethic (a "land ethic") is presented in a convincing manner by Aldo Leopold in *A Sand County Almanac:*

> Conservation is getting nowhere because it is incompatible with our Abrahamic concept of land. We abuse land because we regard it as a commodity belonging to us. When we see land as a community to which we belong, we may begin to use it with love and respect. There is no other way for land to survive the impact of mechanized man, nor for us to reap from it the aesthetic harvest it is capable, under science, of contributing to culture.
>   That land is a community is the basic concept of ecology, but that land is to be loved and respected is an extension of ethics. That land yields a cultural harvest is a fact long known, but latterly often forgotten.[2]

Assuming that we are past the issue of motivation for environmental improvement, what sort of improvement techniques can be used? This, of course, depends very much on the type of environmental problem. For a pollutant, "the solution to pollution is dilution" has been a favorite approach, exemplified by tall smokestacks at power plants. The dilution approach is quite legitimate, because natural systems have the ability to assimilate pollutants of some types up to certain levels of concentration. However, it works only if the problem is small on a global level. Many of our problems no longer are.

The opposite approach is removal of the pollutant at the source. This works well, when it is possible. However, it cannot be assumed that all pollutants can be removed, even in principle. For example, when coal is burned, sulfur dioxide is given off because coal contains a few percent sulfur. This can be viewed as an impurity which can, in principle, be removed. But the burning of coal also produces carbon dioxide. Because oxidation of carbon is the basic process of obtaining energy from coal, carbon dioxide cannot be viewed as an impurity. If you want to get energy from coal, you are, unavoidably, going to produce carbon dioxide. Similarly, energy cannot be extracted from uranium without producing radioactive materials, nor can energy be extracted from any stored-energy source without adding heat to the earth's environment. It is important to stress the unavoidability of these results. Using fossil fuels, uranium, or stored-energy sources necessarily results in carbon dioxide, radioactive material, and extra heat, respectively. There is no technological trick that can avoid these conclusions. (The radioactive material can be removed, stored, and isolated from the environment; the carbon dioxide and heat cannot.)

It is important to realize that certain problems have *no technical solutions,* and that we must look to other areas for answers. Wiesner and York[3] argue cogently that the arms race is such a problem, and that a political solution is required. Hardin[4] argues that population growth is another such problem, requiring changes in human values. While technology and science can make important contributions to the solutions of certain environmental problems, there are other environmental problems of the "no technical solution" variety that require changes in human values and behavior.

Solutions to energy-related environmental problems must take cognizance of the fact that human energy use is growing rapidly, doubling about every 20 years. To hold even, solutions must improve by a factor of 2 every 20 years, or must include a slowing of the energy growth rate.

6.4 Institutions and Laws

Environmental problems are dealt with at the federal, state, and local levels of government by a wide variety of laws and a large number of different agencies, departments, etc. The situation has been in a state of flux over the past decade, and even more rapid changes can be predicted for the coming decade. Therefore, we will omit any discussion of local and state activities and mention only the basic features at the federal level.

The National Environmental Policy Act of 1969 (NEPA) is assuredly a landmark piece of legislation. Title I of this act, included here as Appendix D, deals with a national environmental policy. Section 101 states the policy in fairly general terms, while Section 102 indicates some ways for implementing the policy. To date, the "muscle" of NEPA has been item (2–C) of Section 102, which requires all agencies of the federal government to consider the environmental consequences of their actions and to obtain the views of other agencies and interested parties prior to taking the action. The procedures for doing this are spelled out, namely, an *environmental impact statement* must be prepared, made available for comments, and must accompany the recommendation for the proposed action. The adequacy of an environmental impact statement may be challenged in the courts, and to date the courts have interpreted NEPA as requiring very detailed and careful impact statements. The effect has been that federal agencies that previously had viewed their responsibilities as unrelated to the environment have started thinking about the environment, and have added personnel with competence in that area.

Title II of NEPA creates the Council on Environmental Quality (CEQ), a body which advises the President on environmental matters, reviews and coordinates the environmental impact and environmental control activities of all federal agencies, prepares an annual report on the environment,[5] and generally tries to obtain and make available to the President and the public the "big picture" on the environment.

The Environmental Protection Agency (EPA) is the principal action-oriented environmental institution of the federal government. It administers federal programs dealing with air pollution, water pollution, solid waste disposal, and pesticide regulation. It establishes and enforces standards, monitors and analyzes the environment, conducts research and demonstrations, and assists state and local government agencies.

Until 1975, the Atomic Energy Commission (AEC) had responsibility for the field of nuclear energy for both military and peaceful uses. It conducted and sponsored research and set and enforced standards for operation of nuclear facilities. Its role as both a promoter and regulator of nuclear energy use led to charges of conflict of interest between its two roles. Prior to NEPA, its procedures for licensing nuclear reactors took into account only radiological environmental problems. As interpreted by the courts in the Calvert Cliffs decision, NEPA required the AEC to concern itself with all environmental aspects of nuclear power plants. With the EPA's creation in 1970, the responsibility for setting general radiological standards was transferred from the AEC to EPA. But the AEC still retained responsibility for standards concerning nuclear installations.

In the fall of 1974, there was a reorganization of the federal agencies that deal with energy. The Energy Research and Development Administration (ERDA) was created; it subsumes the AEC's energy research and development programs, the Department of Interior's Office of Coal Research, The National Science Foundation's solar energy program, and other smaller programs. The Nuclear Regulatory Commission (NRC) was also created to take over the regulatory functions of the AEC. It is too early to tell how ERDA and NRC will compare with their predecessors.

A wide array of other federal agencies, bureaus, departments, etc. make decisions and carry on activities relevant to the environment and to energy-related environmental problems. These institutions do not have conservation of the environment as one of their primary missions, and many of them could be regarded as part of the problem rather than part of the solution.

PART B / THERMAL POLLUTION

Thermal pollution is the addition to the environment, as a result of human activities, of heat which under natural conditions would not have been added to the environment. A *redistribution* of heat can also be considered as thermal pollution. (In a more restricted use of the term, thermal pollution is the overheating of bodies of water because of human heat inputs.)

Consider the burning of some form of stored energy such as coal in a power plant, producing electricity which is subsequently consumed in a city some distance away. About 60 percent of the energy content of the coal is added to the environment as heat near the power plant—some directly to the atmosphere via the smokestacks, most to a body of water via the cooling water. The 40 percent of the fuel's energy content that is converted to electricity is also added to the environment as heat, in the city where the electricity is consumed. In the absence of this human activity, the energy would have remained stored in the coal underground.

Alternately, consider the production of electricity from a continuous energy source such as hydropower. Again, heat energy is added to the environment at the point where the electricity is consumed. However, in the absence of human activity, the same amount of heat energy would have been added to the environment at the point where the gravitational potential energy of the water power resource was converted to turbulent motion of water, and then to random molecular motion. Similarly, wind power and tidal energy will be dissipated as heat added to the environment, and the decay of plants returns photosynthetic energy. It is a characteristic feature of all continuous energy sources that their utilization by man does *not* change the amount of heat added to the environment. (It can change the spatial distribution, however.) But human use of *stored* energy sources *does* increase the amount of heat added to the environment.

6.5 Large-Scale Effects—Climate

The amount of heat added per unit area necessary to cause ill effects, and the nature of the ill effects, will vary depending on the size of the area under consideration. In this section we will consider large areas—the earth as a whole, continents, and significant fractions of continents. The ill effect of concern is modification of climate. It was pointed out in Section 2.7 that there are positive feedback effects at work in the earth's energy balance, which could amplify changes caused by man and lead to major climatic changes. There are no universally accepted figures for the amount of heat necessary to be added to the earth to cause such major changes. Estimates are in the neighborhood of one percent of the solar energy input. A definitive experiment has not been conducted . . . yet.

How much will a one-percent change in heat input change the earth's surface temperature? Neglecting feedback mechanisms, a one-percent increase would lead to a one-percent increase in T^4 and a 0.25-percent increase in T. This percentage increase is appropriate for the earth's upper atmosphere, which radiates directly to outer space. For the earth's surface temperature, feedback mechanisms must be taken into consideration. Specifically, a surface-temperature increase (assuming constant relative humidity) will increase the amount of water vapor in the atmosphere, leading to an enhanced greenhouse effect, and an increased temperature difference between the earth's surface and the upper atmosphere necessary to maintain the same outward flow of energy. Empirical data[6] suggest a factor of 2 enhancement due to the feedback mechanism, so human heat input of one percent of the solar value would lead to a surface temperature increase of 0.5 percent, or 1.5°C. This is a large temperature increase. If sustained, it would cause melting of a large part of the polar ice caps, and major climatic changes would ensue.

At present, human use of stored energy is at a rate near 6×10^{12} watts. With $173,000 \times 10^{12}$ watts of solar energy incident on the earth, and an albedo of 0.3, human use represents 0.005 percent of the solar heat input. Human energy use can probably rise one order of magnitude, but not two, without causing problems with global climate. At the current growth rate, human energy use will increase one order of magnitude in about 70 years. If one takes as a goal to bring the worldwide average per-capita energy use up to the present U.S. per-capita level (it is currently one-fifth of that value), and further assumes that the world population can be stabilized at twice its present number (an optimistic assumption in view of current worldwide population trends), then an increase in energy use of a factor of ten is required. Thus we see that global thermal pollution is not an imminent problem, but is one that must be taken into consideration in any long-term thinking.

On a continental scale, a larger energy per unit area is required to cause major climatic changes. The present energy use levels for several regions are listed in Table 6.2. These numbers should be compared with a solar heat input (to the earth's surface and atmosphere) of 250 watts/m². We see that there are large areas, specifically the eastern United States and western Europe, where the human heat inputs approach 1 percent of solar inputs. An order of magnitude increase in energy use in these areas would likely lead to changes in heat convection patterns, affecting climate significantly.

On a smaller geographical scale, cities and industrialized areas have energy inputs from a few percent to a few times solar heat inputs. These areas *do* have climates that differ from their surroundings. No serious climatic problems result because the areas involved are small. One area that does bear watching is Bos-Wash, the megalopolis extending from Boston, Massachusetts to Washington, D.C. It is growing both in area and in energy use per unit area, and is already quite sizeable in both respects.

Table 6.2 Man-generated energy flows in various regions.

	Area (10^6 km²)	Man-made energy flux (watts/m²)
Global average	500.0	0.016
Land surfaces	150.0	0.054
United States	7.8	0.24
Eastern United States (14 states)	0.9	1.1
USSR	22.4	0.05
Central Russia	0.25	0.85
Central Western Europe	1.7	0.74
West Germany	0.25	1.36
Area (10^3 km²)		
Bos-Wash	87.0	4.4
Los Angeles County	10.0	7.5
Nordheim-Westfalen	34.0	4.2
Moscow	0.88	127.0
Manhattan	0.06	630.0

Source: SMIC, *Inadvertent Climate Modification*. Report of the Study of Man's Impact on Climate. Cambridge, Mass.: MIT Press, 1971, pp. 57–59. By permission.

6.6   Local Effects—Water

Point sources of heat energy (factories, power plants, cities) dissipate their heat both to the atmosphere and to nearby bodies of water. As we have just seen, the former may cause minor local climatic changes. The latter is more serious, and is what is usually intended by the term "thermal pollution," that is, the over-heating of lakes, rivers, and other bodies of water.

About 80 percent of the heat dumped into bodies of water comes from steam-electric generating plants. Modern fossil-fuel plants convert 40 percent of the fuel energy to electricity, send 12 percent up the smokestack, and dump the rest, 48 percent, to the cooling water. Nuclear plants of the PWR and BWR variety obtain 33 percent conversion to electricity, with all of the remaining 67 percent going to the cooling water. Thus, per unit of electrical energy generated, modern fossil-fuel plants contribute 1.2 units of aquatic thermal pollution, while nuclear plants contribute 2.0 units. Older fossil-fuel plants have lower efficiencies than the newer ones, with correspondingly more thermal pollution per unit of electricity produced.

Thermal pollution exerts its damage by upsetting or modifying aquatic eco-systems. The harm from this can be economic, psychological/aesthetic, or "nature only," depending on what use was being made of the ecosystem in question. Healthy, productive aquatic ecosystems have evolved in all kinds of water temperatures, from tropical (Panama Lakes) to subarctic (Lake Superior), and in both fresh water and salt water. These ecosystems can be upset by either an increase or decrease in temperature. While most of our comments will be about temperature increases, since this is what human activity usually causes, it is important to realize that warm water is not inherently worse than cold water, nor are temperature increases inherently worse than temperature decreases. Temperature changes cause ecosystem disturbances in a variety of ways. (1) Large temperature increases (or decreases) can directly kill many aquatic species. Abrupt changes in temperature are more hazardous than gradual ones. For many species of fish, the lethal temperature is 10° to 15°F above their customary living temperature. (2) Under some circumstances,[7] oxygen is in short supply in rivers and lakes, and increased temperatures aggravate the situation considerably. The solubility of oxygen in water decreases with increasing temperature (14.5 ppm at 0°C; 10 ppm at 15°C; 7.6 ppm at 30°C). The hemoglobin in a fish's blood had reduced affinity for oxygen at higher temperatures, and therefore the fish's oxygen delivery system becomes less efficient. The metabolic activity of most species of fish increases with increasing temperature (roughly doubling for every 10°C increase), and the need for oxygen therefore rises. Finally, if the lake or river contains a load of organic matter in the process of decaying (e.g., sewage), increased temperatures will speed the rate of decay, and consequently the rate at which oxygen is used in the decay process. (Again, let me stress that warm water is not inherently worse than cold water. Rather, an increase in temperature above "normal," in conjunction with a dissolved oxygen concentration already below "normal," can result in oxygen supply problems for those species adapted to "normal" conditions.) (3) Changed rates of biological activity (caused by temperature changes) can cause problems other than that of oxygen supply just discussed by changing the rate of growth and the life span of various species. For example, rapid growth of algae or pond weeds, stimulated by increased temperatures, is often undesirable. (4) The resistance to disease of many species of fish decreases as the temperature rises above their "preferred" temperature. (5) The behavior patterns of many species are keyed to temperature and temperature changes. Breeding and migratory behavior can be upset by "artificial" temperature changes. For example, hot effluents into the Hudson River are regulated so they do not reach all the way across the river, because if they did, they would form a "thermal fence" to the normal migration of fish. (6) Each species of plant or animal has a "preferred" range of temperature (often fairly narrow) in which it functions best. A spatially localized thermal input to a healthy lake or river (no oxygen shortage) is usually well tolerated, because fish can move around and find their own preferred temperature. On the other hand, even a small temperature change, if it involves a large fraction of an ecosystem (e.g., by area or by volume) and if it is sustained for a long period of time (years), can give a competitive advantage to a different species and result in a change in species composition of the ecosystem. For example, cold-water fish such as salmon and trout would give way to warm-water fish such as perch and bass. Similarly, as temperatures rise above 77°F, diatoms give way to green algae, which at 95°F give way to blue-green algae.

Needless to say, the various species in an ecosystem are extremely interdependent, and so damage to one component (e.g., a change in the type of algae) can cause changes in some other component seemingly far removed (e.g., by altering the nature of the food chain). If the temperature of some aquatic ecosystem is increased and maintained at a new temperature (but not above 95°F, a lethal temperature for almost all fish), a new ecosystem will develop in time. It may not be as "desirable" as the one it succeeded, but it will be functioning and healthy. However, if the temperature is allowed to vary, as occurs when a power plant shuts down for repairs, a stable ecosystem cannot develop.

How much temperature increase can an aquatic ecosystem tolerate? From the preceding discussion it is evident that the answer to this question will vary from one ecosystem to another. For most lakes and rivers, a long-term temperature rise of 10°F or more would cause major changes. How adverse they would be would again depend on the particular ecosystem. It has been suggested that temperature increases in the 3° to 5°F range in many cases will not cause significant changes. A fairly conservative requirement might be that increases must not exceed 2°F, or 1°C. If the temperature increase involves only a small fraction of an ecosystem (by area), then a larger increase can be tolerated.

Under the Federal Water Pollution Control Act Amendments of 1965 and 1972, aquatic thermal pollution standards are set by the individual states as part of their water quality standards. These standards must be reviewed and approved by EPA. Typical standards allow temperature increases of less than 5°F for rivers and 3°F for lakes, with more restrictive limits for particularly sensitive ecosystems.

6.7   Single-Pass Cooling Using Rivers

The standard practice for disposing of waste heat from a power plant is *once-through cooling*. Water is drawn from a river, lake, or estuary, is passed through the steam condenser heat exchanger where its temperature rises by perhaps 10°C, and is then returned to the body of water from which it was drawn. In this section, we will consider the adequacy of this approach using rivers; in the following section we will consider this approach using lakes.

The present U.S. rate of waste heat from generation of electricity is in the neighborhood 0.3×10^{12} watts. The average discharge of all U.S. streams into the ocean is 2×10^6 cu ft/sec. If heat is added at the rate of 0.3×10^{12} watts to a stream flow of 2×10^6 cu ft/sec, then the water temperature will increase by 1.4°C. This means that if all U.S. stream flow is used, but used only once, and if rivers supply all the cooling water, then on the average river, temperatures will be raised by 1.4°C. (In practice, some fraction of a river's flow would be used, raised in temperature by perhaps 10°C, and then mixed with the rest of the river flow, with the "recombined" river having an average increase of 1.4°C.) But "average discharge" is an annual average obtained from high flows and low flows; a more useful number would be the typical annual low flow, which is about a factor of two smaller. Then under the above assumptions, on the day of annual low flow, the average temperature increase would be 2.8°C, and those rivers whose flow has dropped more than average would suffer still higher increases. These increases would be too large, a factor of three above our suggested figure of 1°C, and right at the limit of 5°F proscribed by many states.

Stream flow can be used more than once, however. After being used the first time, the river is allowed to cool down as it flows some distance, and then it is used again. In this context, cooling down means transferring heat from the river to the atmosphere by radiation, conduction, and evaporation. Typically, half of the heat is transferred by evaporation. If half of 0.3×10^{12} watts is dissipated by evaporation, water is evaporated at a rate of 2500 cu ft/sec, or 0.12 percent of the average discharge to the ocean. This is a *consumptive use* of water, i.e., one that makes the water unavailable for other uses on that trip through the hydrological cycle.

By using stream flow an average of three times, present power-plant cooling requirements could be handled by once-through cooling using rivers, keeping temperature increases to 1°C. The number of times stream flow can be reused is determined by the requirement that it cool down between uses. When a river is being utilized in this way, it is functioning more like a series of lakes (delivering heat to the atmosphere) rather than like a river (delivering heat to the ocean). The heat delivered to the atmosphere is proportional to the surface area of the river between power plants times the temperature increase of the river (see Section 6.8 and Problem 6.3). Studies for the Ohio River[8] suggest a plant spacing of 25 miles. Allowing for the lower flow as one moves upstream, the 1000-mile stretch of river was found to be capable of supplying 15 times more cooling capacity than a single use of the stream flow at the river mouth. The use factor of 15 is probably appropriate for the Mississippi River drainage basin, but is too high for the east and west coasts where rivers are much shorter. For a very rough estimate, we assume a use factor of 6 for the United States as a whole. In other words, maximum reuse of stream flow allows an increase of cooling capacity a factor of 6 over single use. Stream flow in the United States is therefore capable of supplying twice the total present power-plant cooling requirement with a 1°C permitted rise, and 6 times the present requirements with a 3°C permitted rise. Problems of geography (rivers not being where the power plants are needed) and flow variations in excess of average will lower the actual usable cooling capacity.

How big a river is needed for a modern power plant? A typical new nuclear plant will generate 1000 MW$_e$ of electrical power, dumping 2000 MW of heat. (A modern fossil-fuel plant generating 1000 MW$_e$ will dump 1200 MW of heat.) For a 1°C rise, the nuclear plant requires 17,000 cu ft/sec of water, while the fossil-fuel plant requires 10,000 cu ft/sec. Clearly, a very large river is required to handle a large power plant. The flows of the larger U.S. rivers are listed in Table 6.3.

6.8   Single-Pass Cooling Using Lakes

Heat added to a lake by once-through cooling is transferred to the atmosphere by conduction, radiation, and evaporation. (A small amount of heat may be carried out of a lake along with water flowing out.) Before considering the processes in detail, we will use our "one-percent law" to obtain a rough estimate of the

Table 6.3   Large rivers in the United States.

River	Average discharge (10^3 cu ft/sec)	Length (miles)
Mississippi	640.0	3710
Columbia	262.0	1243
Ohio	258.0	1306
St. Lawrence	243.0	—
Yukon	240.0	1770
Atchafalaya	183.0	135
Missouri	76.3	2533
Tennessee	64.1	900
Red	62.3	1270
Kuskokwim	62.0	680
Mobile	61.4	780
Snake	50.0	1038
Arkansas	45.1	1450
Copper	43.0	280
Tanana	41.0	620
Susitna	40.0	300
Susquehanna	37.2	444
Willamette	35.7	270
Alabama	32.4	735
White	32.1	720
Wabash	30.4	529
Pend Oreille	29.9	490
Tombigbee	27.3	525
Cumberland	26.9	720
Stikine	26.0	310
Appalachicola	24.7	524
Sacramento	23.0	377
Illinois	22.8	420
Koyukuk	22.0	470
Porcupine	20.0	460
Hudson	19.5	306
Allegheny	19.3	325
Delaware	17.2	390

Source: Copyright 1973, *The World Almanac and Book of Facts 1974*, p. 438.

cooling capacity of lakes. In this application, the "one-percent law" suggests that man-induced heat additions to a lake should be kept below one percent of the natural heat additions for the lake as a whole. Considering a lake and the atmosphere above it, solar energy is being added at a rate averaging 250 watts/m², suggesting that man-induced energy additions be kept below 2.5 watts for each square meter of lake surface area.

A 1000-MW_e nuclear power plant, with 2000 MW of waste heat, thus requires a lake surface of 0.8×10^9 m², namely, 800 km² or 300 sq mi. That is a big lake, as can be seen from Table 6.4 which lists the larger lakes and reservoirs in the continental United States. The Great Lakes each can tolerate several large power plants, while few other lakes can tolerate any.

Table 6.4 Lakes and reservoirs in the continental United States with areas larger than 200 sq mi. (Some lakes in Florida and Minnesota are not listed.)

Lake/reservoir	Area (sq mi)	Mamimum depth (ft)
Superior	31,800	1333
Huron	23,000	750
Michigan	22,400	923
Erie	9,900	210
Ontario	7,600	802
Great Salt (Utah)	1,500	48
Red Lake (Minnesota)	751	35
Okeechobee (Florida)	700	20
Pontchartrain (Louisiana)	621	15
Garrison Reservoir (North Dakota)[a]	609	180
Oahe Reservoir (North Dakota–South Dakota)[a]	580	200
Champlain (New York–Vermont)	490	400
Fort Peck Reservoir (Montana)[a]	386	220
Salton Sea (California)	360	48
Powell (Arizona–Utah)[a]	252	580
Kentucky (Kentucky–Tennessee)[a]	250	60
Mead (Nevada–Arizona)[a]	247	432
Winnebago (Wisconsin)	215	22

[a] Man-made.

Source: Copyright 1973, *The World Almanac and Book of Facts 1974*, p. 439.

Now let's look at the energy transfer processes. The energy exchanged between a lake and the atmosphere (per unit area and time) can be written

$$E_{ex} = E_{IR} + E_c + E_e - E_s. \tag{6.1}$$

E_s is the radiant energy (both from the sun and from the atmosphere) absorbed by the lake. It is independent of the lake's surface temperature T_s. E_{IR} is the infrared energy radiated by the lake. Since water is essentially black in the infrared, it can be written $E_{IR} = \sigma T_s^4$. E_c is the heat transferred by conduction from the lake to the atmosphere. It is proportional to $T_s - T_a$, where T_a is the temperature of the atmosphere. E_e is the energy transferred to the atmosphere by evaporation and is proportional to the difference between $p_s(T_s)$, the vapor pressure of water at the temperature T_s, and p_a, the partial pressure of water vapor in the atmosphere.

For both E_c and E_e, the constants of proportionality will depend on wind, which carries water vapor and sensible heat away from the air/water interface, "making room" for further evaporation and conduction. It is reasonable to expect that the dependence on wind will be the same for both terms. We therefore write:

$$E_e = (p_s - p_a)f(w), \tag{6.2}$$

$$E_c = (T_s - T_a)\,\kappa\, f(w), \tag{6.3}$$

where $f(w)$ is the "wind speed function," large when the wind is strong and small when the wind is weak, and κ is a constant. Both $f(w)$ and κ can be determined empirically.[9]

E_{ex} need not equal zero, since on any given day a lake can either gain or lose energy. However, for a lake with no man-induced energy additions, the annual average of E_{ex} must be zero. If an energy E_{man} is added to the lake by human activities, then E_{ex} will rise until its annual average equals the annual average of E_{man}, i.e., until the lake delivers to the atmosphere a net amount of energy equal to that given to it by human activities. For this to happen, the surface temperature of the lake must increase by ΔT_{man}:

$$E_{man} = \left(\frac{dE_{ex}}{dT_s}\right) \Delta T_{man}. \tag{6.4}$$

For small lakes (all lakes in the United States other than the Great Lakes), the temperature and humidity of the atmosphere (T_a and p_a) are independent of the surface temperature of the lake T_s, and the derivative can be written

$$\frac{dE_{ex}}{dT_s} = 4\sigma T_s^3 + \left(\frac{dp_s}{dT_s} + \kappa\right)f(w). \tag{6.5}$$

For the Great Lakes, T_a and p_a increase as T_s increases, reducing dE_{ex}/dT_s below its value given in Eq. 6.5 by about 30 percent.

Asbury[9] has performed calculations applicable to Lake Michigan, and finds for dE_{ex}/dT_s that (a) there is a factor of two seasonal variation due to variations in water temperature and wind speed, (b) for an annual average, the evaporation term is 50 percent of the total term, the radiation term 20 percent, and the conduction term 30 percent, and (c) the annual mean value is 22 watts/m²-°C. Thus a man-caused increase of 2.5 watts/m² (1 percent of solar energy flux) would cause an annual mean surface temperature increase of 0.11°C.

We must consider subsurface temperatures also. Most lakes of any appreciable depth in northern latitudes exhibit a property known as *thermal stratification* during the summer months, due to the following facts. (a) Down to 4°C, cold water is more dense than warm water. (b) Water conducts heat poorly.

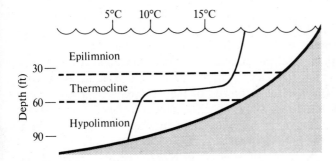

Fig. 6.2 Temperature profile of a thermally stratified lake.

(c) A lake is heated and cooled from the air water interface, i.e., from the top of the lake. Figure 6.2 shows the temperature profile of a typical thermally stratified lake. Heat added to the lake warms the upper region (*epilimnion*) which, as a result, becomes lighter than the lower region (*hypolimnion*) and therefore floats on it. There is a fairly sharp boundary (*thermocline*) between the two regions and they do not mix. In the fall and winter as the lake cools, the surface water becomes colder than the underlying water and sinks, cooling the epilimnion. When the epilimnion density approaches that of the hypolimnion, wind action causes vertical mixing and thermal stratification ends. The lake assumes a uniform temperature and cools toward 4°C. If further cooling occurs, the colder water may float, perhaps eventually forming ice. In the spring as the surface warms, there is vertical mixing until lake temperatures reach somewhere above 4°C. With further warming, the surface waters float on the more dense underlying waters, and thermal stratification develops.

During the period of thermal stratification—typically May to November— the hypolimnion makes no contact with the atmosphere. Its dissolved oxygen supply is therefore not replenished. (The amount of oxygen reaching the hypolimnion by diffusion across the thermocline is negligible—see Problem 6.6.) Should the drain on the dissolved oxygen be too large, the oxygen in the hypolimnion may be exhausted before autumn when vertical mixing resumes. This will of course have adverse effects on many species of animals living in the hypolimnion.

The water returning to a lake from a power plant is usually warmer than the water at the surface of the lake and may therefore float on top of it, forming yet another layer, the *thermal plume*. This layer is separated from the epilimnion by a *microthermocline,* just as the epilimnion is separated from the hypolimnion by a thermocline. The thermal plume loses energy to the atmosphere and cools as it moves away from the power plant. When its temperature gets sufficiently close to that of the epilimnion (a few degrees, depending upon strength of wind and wave action), vertical mixing occurs and the thermal plume disappears. The thermal plume thus delivers part of the waste heat directly to the atmosphere and the rest to the epilimnion, which in turn delivers it to the atmosphere. Near the power plant, the thermal plume is several feet deep and has a temperature near that of the power plant return water, which may be 10°C above that of the epilimnion. Away from the power plant near the outer reaches of the plume, its depth is less than a foot and its temperature only a few degrees above that of the epilimnion. The temperature of the hypolimnion is not changed by the heat addition from the power plant, and the epilimnion temperature increase is smaller than the surface temperature increase. (Often when waste heat is added to a *river,* the warm return water will float on top of the remaining river water. The discussion in Section 6.7 did not allow for this. The consequence is that the surface water temperature increase is greater than indicated by the considerations of Section 6.7, and the bottom water temperature increase is less.)

An informative case study is the proposal by New York State Electric and Gas Corporation to build an 830 MW_e nuclear plant (Bell station) on Cayuga Lake, one of the Finger Lakes in central New York State. Cayuga Lake is a moderate-sized (66 sq mi), deep (435 ft maximum, 180 ft mean), cold lake, thermally stratified from May to November. The proposed plant would draw water from the hypolimnion at a depth of 100 ft and return it to the epilimnion, adding heat at a rate of 1600 MW, which is about 10 watts per square meter of lake surface (4 percent of solar energy flux). An existing 300 MW_e fossil-fuel plant (Millikan station) is already adding 2 watts/m² to the lake (1 percent of solar flux). Several groups have studied the effects to be anticipated from the combined operation of the existing and proposed plants. Expectations are that (a) the mean surface temperature would increase by 0.5°C, and the greatest increase would be 1°C, occurring in April, (b) the volume of the epilimnion would increase by 20 percent, and (c) thermal stratification would start earlier and end later, with a total extension of about 10 days. All three of these effects increase the biological activity of the epilimnion, perhaps by a total of 15 percent annually. Dead plant and animal matter settles to the hypolimnion where its decay consumes oxygen. The combined effect of increased biological activity and longer thermal stratification is that the dissolved oxygen level deep in the hypolimnion becomes marginal by the end of the period of stratification. In addition to thermal effects, the new power plant's cooling system would bring nutrients from the bottom waters to the surface, further increasing biological activity. The Corporation's estimate was that the thermal effects would reduce the dissolved oxygen concentration at the 400-ft level by 0.2 ppm, and nutrient pumping would reduce it by another 0.2 ppm, but that dissolved oxygen concentration would not drop below 5.6 ppm. Since concentrations much below 5 ppm are deleterious, it appears that the power plants would use up a significant fraction of the lake's margin of safety. My conclusion is that this case demonstrates that 5 percent of solar energy flux is too large an addition for a high quality recreational/aesthetic lake, and the fact that the plants would satisfy the legal requirement, that the surface temperature increase not exceed 3°F (1.7°C), suggests that the standard is not stringent enough. (The Corporation has currently abandoned plans for a nuclear plant, and instead intends to build a fossil-fuel plant of the same size at the same site. Thermal pollution problems were not given as the reason for the switch.)

Cayuga Lake is an example of a lake whose ability to tolerate waste heat has been reduced by other human abuses. While in no sense grossly polluted, Cayuga Lake receives both fertilizer runoff and treated sewage effluent. Equally important, much of its protective[7] marshlands and woodlands have been filled and cleared, respectively. As a result, even without thermal pollution, its dissolved oxygen content is rather low by the end of the summer. Were Cayuga Lake in pristine condition, it could tolerate thermal additions of 10 percent solar input. However, very few lakes in the United States are in such condition, and thermal standards must be based instead upon typical conditions. Cayuga's condition is, if anything, better than average.

If we maintain the conservative limit of 1 percent solar, which is 300 sq mi per 2000 MW of waste heat, we see that the five Great Lakes can provide a total of 0.6 × 10¹² watts of cooling capacity, which is twice the present level of waste heat from generation of electricity. The total cooling capacity of all other lakes in the United States is much smaller.

6.9   Thermal Pollution Abatement Schemes

In those cases where a lake or river is large enough to safely absorb the heat from a power plant via once-through cooling, care is still required to ensure that the heat is distributed over the entire body of water. The heat must not remain concentrated in a small area so it can cause damage there. This is particularly likely to be a problem in attempts to utilize the full capacity of any of the Great Lakes. With the current techniques, the plume of heated water returning from a power plant to the lake turns under the influence of the Coriolis force and hugs the shoreline, so that the cooling capacity of the center of the lake is not utilized. Care is also required to ensure that the area immediately around the cooling water return point, where the water will unavoidably be several degrees warmer than ambient, is not of an ecologically sensitive nature. For example, breeding areas and/or migration paths must not be upset. The design of an environmentally sound once-through cooling system must be tailored for a particular site and must be based on a thorough ecological study of that site.

As seen in the preceding two sections, a conservative utilization of rivers and lakes for once-through cooling provides only a factor of four more capacity than present needs. This figure is reduced further by geographical distribution because capacity is not always located where it is needed. At present, electrical power use is doubling every ten years, so that continued growth necessitates consideration of alternative cooling schemes.

1. Once-through cooling using the ocean provides essentially unlimited cooling capacity. Its utilization requires the precautions mentioned at the beginning of the section, and materials that will withstand the corrosiveness of salt water must be used. Of course, it is geographically limited to coastal sites.

2. *Cooling ponds* are bodies of water constructed explicitly for cooling purposes. Their utilization is identical to that of once-through cooling with lakes, except that there is no concern about thermal pollution because the pond is not intended to be a functioning natural ecosystem. Heat inputs are in the neighborhood of 200 watts/m^2, and surface temperatures are perhaps 10°C higher than those of natural lakes.

As a variation on the cooling-pond theme, it should be possible to design an artificial ecosystem in which the waste heat is *beneficial*. A system of marshes and ponds could be used for the final stages of sewage treatment, fish farming, boating and swimming, as well as heat disposal. A few attempts in this direction have combined heat disposal with oyster farming, and heat disposal with sewage treatment.

3. *Evaporative* (or *wet*) *cooling towers* transfer heat from the cooling water to the atmosphere by evaporation. This transfer takes place as air flows past the heated cooling water being sprayed or trickled over large surfaces. These cooling towers can be either "natural-draft" (air motion due to convection) or "mechanical-draft" (air motion due to fans).

4. *Dry cooling towers* transfer heat from the cooling water to the atmosphere by conduction as air flows past the heated cooling water contained in tubing.

(The radiator of an automobile functions on exactly the same principle.) Like wet cooling towers, these can be either natural-draft or mechanical-draft.

These abatement schemes have environmental problems of their own. Cooling ponds tie up substantial terrain (4 sq mi for 2000 MW of waste heat) for their specific purpose. Cooling ponds and wet cooling towers represent a consumptive use of water. Towers consume two to three times as much water as cooling ponds or lakes and rivers. There is formation of fog and ice in the neighborhood of cooling towers under some meteorological conditions. Cooling towers are large (see Fig. 6.3) and are therefore an aesthetic blemish on the landscape. The wet natural-draft cooling towers at TVA's Paradise plant in Kentucky are 437 ft high and 320 ft in diameter at the base. For the same cooling capacity, natural-draft towers are larger than mechanical-draft, and dry towers are larger than wet towers. Mechanical-draft towers are quite noisy. Dry towers cannot cool to as low a temperature as wet towers can, which in turn may not attain as low a temperature as once-through cooling. Higher temperatures lower plant efficiency, which increases fuel consumption, air pollution or radioactivity, and waste heat rejection, per unit of electricity generated.

Fig. 6.3 Natural-draft evaporative cooling towers at the TVA's Paradise steam-electric plant in western Kentucky. These three towers are used to supplement the cooling capacity of the Green River. Note trucks in the foreground for scale. (Photo courtesy of the Tennessee Valley Authority.)

All the abatement schemes discussed above refer to local, aquatic thermal problems. The solutions all amount to dilution, using either the atmosphere or the ocean. For the problem of large-scale thermal pollution—continental and global—the only solution is to keep the use of stored energy from growing excessively.

PART C / RADIOACTIVITY

After discussing the nature of ionizing radiation, biological ill effects, sources of radiation, and legal dose limits, we consider the major future source of ionizing radiation—nuclear energy.

6.10   Ionizing Radiation

Energetic charged particles (α-particles and β-particles) from radioactive decays or other sources lose energy as they pass through matter by a process called *ionization*. As the charged particle passes by a molecule, it exerts a strong attractive or repulsive electrical force on the electrons in the molecule. In some cases an electron (e) may be torn away from the molecule (M), thus ionizing the molecule and forming the ion pair M^+ and e^-. The energy lost by the fast charged particle in creating the ion pair is that required to break the electron away from the molecule plus the kinetic energy imparted to the electron. A typical value is in the neighborhood of 30 eV. Thus a 3-MeV alpha-particle will create in the neighborhood of 10^5 ion pairs as it slows down and stops.

How do neutral particles (i.e., γ-rays and neutrons) from radioactive decay or from nuclear reactions lose their energy? They first give a significant fraction of their energy to a charged particle, and the charged particle then loses energy through ionization as described above. The γ-ray can give up part or all of its energy via the photoelectric effect ($\gamma + \text{atom} \rightarrow \text{ion}^+ + e^-$), pair production ($\gamma + \text{nucleus} \rightarrow e^+ + e^- + \text{nucleus}$), or Compton scattering ($\gamma + e \rightarrow \gamma + e$). The neutron gives up its energy only through a nuclear reaction, such as the scattering from a proton ($n + p \rightarrow n + p$).

In subsequent considerations, we will need units for *radioactivity* and for *radiological dose*. In discussing radioactivity, the number of decays per unit time is of interest. The *curie* (abbreviated Ci) of any radioactive substance is defined as that amount of the substance which will gives rise to 3.7×10^{10} decays per second. (One gram of radium is about equal to one curie.)

In discussing radiological dose, the number of ion pairs created per unit volume or per unit mass is relevant. That is, if there is a region of space through which α-particles, β-particles, or γ-rays are passing, and one places an object (e.g., a person) in that region, how much "dose" would the object receive, i.e., how many ion pairs per gram of tissue would be formed? Three different units are in common use. (1) The *roentgen unit* (abbreviated R) is the dose that will create one stat-coulomb of charge in one cubic centimeter of dry air. (2) The *rad* is the dose that will deposit 100 ergs in one gram of material. The roentgen and the rad are about equal. The biological effect of ionizing radiation is largely proportional to the number of ion pairs. However, if the ion pairs are very closely spaced, as occurs with α-particles and neutrons, then there is an enhanced biological effect. For this reason the *roentgen-equivalent-man* (abbreviated rem) is often used. For β- and γ-rays, 1 rem = 1 R; for α-particles, neutrons, and other very densely ionizing particles, the rem is that dose which has the same biological effect on a human being as 1 R of β or γ radiation.

In any given situation, there is a relationship between the amount of radioactivity of some source (number of curies) and the radiological dose received at some point near the source (number of roentgens).

$$\text{dose (R)} = \frac{B \times \text{activity (curies)} \times \text{time}}{\text{distance}^2} \tag{6.6}$$

The number of roentgens is proportional to the number of curies times the duration of the exposure divided by the square of the distance from source to point. The constant of proportionality B depends on the type of particles given off (α, β, γ, etc.) and the energy of the particles. Both of these are a characteristic of a particular radioactive nuclide. If material that absorbs the ionizing radiation is placed between the source and the point in question, the dose is reduced below that indicated by Eq. 6.6. With sufficient amounts of this "shielding," doses can be reduced by very large factors.

Alpha-particles from radioactive decay are typically emitted with energies in the 5-to-10-MeV range. With these energies they ionize very heavily, lose energy quickly, and travel only a short distance. A 10-MeV α-particle will travel about 10 cm in air or 0.015 cm in water or human tissue before stopping. From this it follows that α-emitting radionuclides are hazardous principally if ingested or inhaled. Then they give a very localized but very heavy dose to the region where they lodge. Gamma-rays from radioactive decays have energies in the 0.1 to 5.0 MeV range. The distance they travel is not sharply defined, but average about 300 meters in air and 0.5 meters in water. Thus they need not be emitted internal to the body to be harmful, and cause a diffuse, "whole-body" dose rather than a localized one. One-MeV electrons (typical of β-decay of radioactive nuclei) behave in an intermediate fashion, traveling 3.5 m in air and 0.5 cm in water. Fission-fragment decay chains usually involve emission of γ-rays as well as β-rays. High-atomic-number radioactive elements (radon and plutonium) often emit low-energy γ-rays as well as α-particles.

6.11 Biological Effects of Ionizing Radiation

The normal functions of living cells depend on genetic information which is stored in polymers of *deoxyribonucleic acid* (abbreviated DNA). Ionizing radiation may cause damage either by directly ionizing an atom of the DNA or by ionizing a nearby molecule so that the resulting ion reacts chemically with the DNA. The chemical structure of DNA is a double-stranded, linear polymer (the double helix) composed of four kinds of monomers symbolized by the letters A, G, C, and T. In biochemical terms, the genetic message is spelled out by the precise sequence of monomers in a strand—it is replicated when the cell replicates, and it is read out ("transcribed" and "translated") to specify the chemical composition of every protein molecule in the cell. In biological terms, the DNA of living cells is packaged into protein-coated units called *chromosomes,* and these in turn are divided into functional units called *genes.* Each gene corresponds to a long DNA polymer—long enough to specify an entire protein molecule—and the function of the gene can be disrupted by the loss or replacement of even one atom of the DNA strand. It takes a special chain of events to produce so large an effect from so small a cause. First, the change must cause a misreading (the altered A, for example, would be mistaken for a G) which will, in turn, change the composition of one protein produced in the cell. Second, the change must be in a place that controls the function of the protein. Third, the defect in the protein must show up as a defect in a function of the living cell—a *deleterious mutation.* When a deleterious mutation occurs in a reproductive cell and the damage kills the resulting offspring, it is called a *lethal mutation.* The more disruptions that occur (by ionization, removal, or replacement of atoms in the DNA molecule), the more the likelihood of deleterious mutations or even of mechanical breaks in the chromosomes which lead to the loss of whole clusters of genes when cells divide.

Living organisms have many lines of defense against mutations. Each cell has special chemical equipment to quench some of the radiation-excited molecules and ions before they damage the DNA. Second, each chromosome has built-in repair mechanisms that detect and remove damaged DNA sequences by enzyme action, and repair breaks in DNA strands. Often though not always, mutations can be returned to normal in this way. In multicellular organisms, mutated and defective cells are replaced with normal new ones (though not in all tissues). And finally, the most advanced organisms such as the vertebrates have immune mechanisms to detect and remove some of the cells that were changed by mutation but survived.

In summary, the amount of radiation damage depends on the interaction between ionization reactions that disrupt the DNA in cells and the biological processes that correct some of the resulting damages.

Three types of damage to the organism as a whole due to ionizing radiation may result.

1. *Acute somatic damage* results when a large dose of radiation is sustained. It causes fatal damage to a large number of cells, resulting in *radiation sickness* (nausea, vomiting, headaches, weakness), and sometimes death. The time scale is hours to days. In human beings, a dose of 100 rem will cause mild radiation sickness; 250 rem causes severe radiation sickness and occasionally death; 500 rem has a 50-percent chance of causing death; 800 rem is almost sure to cause death.

2. *Delayed somatic damage* can result when an organism receives a dose that

is not fatal to the organism. Those cells that were lethally damaged will not re-produce and will be eliminated. Those cells that are damaged nonlethally will stay with the organism, but may cause malfunctions later on. These malfunctions include cancer, cataracts, prenatal abnormalities, and nonspecific shortening of lifespan. Because the delayed somatic phenomena are not of a yes/no variety, but rather are manifest as an increased probability of occurrence, and because damages are not different in kind from the diseases that show up with aging, the relation of dose to the resulting damage can be expressed as speeded-up aging. The relation between the shortening of life and the dose of radiation was carefully measured in mice for relatively large doses, and is approximately known for people. There is still controversy about the effect of small doses. Does the amount of damage approach zero just as the dose approaches zero, or does the exposed population reach the longest lifespan (that is, zero life shortening) at some small but measurable dose of radiation? The controversy is over the existence of a threshold for damage, which controversy also appears in estimates of genetic damage. Rough estimates are that the probability of an individual contracting cancer is doubled by a dose of 100 rem, and that a 1-rem dose has a 2×10^{-4} chance of causing cancer. Note that for acute effects, the dose must be delivered all at once, while for delayed effects it may be spread out in time.

3. *Genetic damage* results when a reproductive cell (sperm or egg) is non-lethally damaged. A reproductive cell that is lethally damaged dies, causing no harm, but one that is nonlethally damaged may give rise to a genetically defective offspring. If the defect is of a severe nature, that offspring may die before repro-ducing, thus ending the damage. The "less severe" genetic defects are passed on to subsequent generations, perhaps causing more total misery. The damage that these less severe defects bring about is measured as an increase in the mutation pressure on the population. Every living population lives under considerable mu-tation pressure and can tolerate a large increase in this pressure without becoming extinct. In the long run, mutation pressure is even an advantage to continuing life—evolutionary progress is the outcome of a creative interplay of mutation pressures and selection pressures. However, in the short run of only a few genera-tions rather than a few hundred thousand, mutation pressure only adds to the load of defective births and hereditary ills. To the extent that any radiation-exposed human being may later become a parent, the exposures add to the mu-tation pressure on the entire human population to which he or she belongs.

6.12   Sources and Legal Limits of Radiation Exposure

Sources of radiation exposure in the United States are listed in Table 6.5. There is considerable variation in the natural sources. The dose from cosmic rays in-creases with elevation, being about twice as large in Denver at 5000 ft as at sea level. The earth's radioactivity depends on soil and rock composition, and varies from place to place. Of the man-made sources, diagnostic medical x-rays are by far the largest. They could and should be substantially reduced by better collima-

Table 6.5   Radiation exposures in the United States.[a]

	Millirems/yr[a, b]
Natural sources	
A. External to the body	
1. From cosmic radiation	50.0
2. From the earth	47.0
3. From building materials	3.0
B. Inside the body	
1. Inhalation of air	5.0
2. Elements found naturally in human tissues	21.0
Total, natural sources	126.0
Man-made sources	
A. Medical procedures	
1. Diagnostic x-rays	50.0 (up to 95 in 1970)
2. Radiotherapy x-rays, radioisotopes	10.0
3. Internal diagnosis, therapy	1.0
Subtotal	61.0
B. Atomic energy industry, laboratories	0.2
C. Luminous watch dials, television tubes, radioactive industrial wastes, etc.	2.0
D. Radioactive fallout	4.0 (down to 1.0 in 1970)
Subtotal	6.2
Total, man-made sources	67.2
Overall total	193.2

[a] Estimated average exposures to the gonads based on the 1963 report of the Federal Radiation Council.
[b] One millirem equals 10^{-3} rem.
Source: David R. Inglis, *Nuclear Energy: Its Physics and Its Social Challenge*. Reading, Mass.: Addison-Wesley, 1973, p. 130. By permission.

tion of the x-ray beam, better shielding of those parts of the body not being x-rayed, and faster film. The contribution of radioactive fallout from nuclear weapons tests has been decreasing since above-ground testing was curtailed in 1963. Radiation attributable to the nuclear energy industry appears negligible in Table 6.5. However, the industry is expected to grow by a factor of 100 in the next few decades. Furthermore, even now there exists the *potential* for a large dose if some accident occurs. We should therefore take a close look at nuclear energy now, before the growth occurs.

Various national and international agencies set guidelines on limits to radiation dose for radiation workers and for the general public. Until recently, the AEC had the authority to set legal limits, and generally followed the guidelines set by the Federal Radiation Council. Upon its creation, the EPA assumed the function of the Federal Radiation Council and was given responsibility for setting general standards. The AEC retained the task of translating the general standards into rules applicable to the nuclear-power-industry facilities. The precise dividing line between the responsibilities of EPA and AEC is somewhat unclear and has recently (1973) been reinterpreted to give most of the responsibility back to AEC. Presumably, these responsibilities now belong to NRC (see Section 6.4).

Current standards for permissible doses of radiation are the following.

Radiation workers—less than 3.0 rem in any consecutive 13 weeks; long-term accumulation to age n years is not to exceed $5(n - 18)$ rem.

Public $\begin{cases} \text{Occasionally exposed individual—less than 0.5 rem in any one year} \\ \text{General public—less than 0.17 rem in any one year} \end{cases}$

These limits do not include natural radiation or medical radiation. The rationale behind allowing employees at nuclear facilities a higher dose than the general public is that an average dose of 5 rem per year, while not completely negligible from the point of view of somatic damage, is small compared to other typical industrial hazards. For genetic damage, the danger is not to the individual but to the population as a whole. What matters is the total dose in man-rem, whether it comes from a high dose to a small number of individuals or a small dose to a large number of individuals. Since radiation workers constitute a very small fraction of the population, the higher dose permitted them makes little contribution to genetic damage to the population.

There has been criticism of these limits, and it has been argued that an exposure of the general public to 0.17 rem per year would lead to unacceptable health damages. A conservative approach would be to apply the "one-percent law" here—to limit average exposures to the general public to 1 percent of the natural level. At first sight, this appears to require a 100-fold reduction in the limit of 0.17 rem per year. However, radiation from nuclear facilities is not uniformly distributed, and in assuring that the dose to individuals in nearby communities is below 0.17 rem/yr, the dose to the average individual, nationwide, will be kept much smaller.

Since June 1971, the AEC has required, for light water reactors (LWR) only, that the limits be 5 percent of natural radiation for the occasional individual and 1 percent for the general public. This reduction in the limits by a factor of 100 was made because LWR's could meet the new limits without undue strain. These lower limits do not apply as yet to fuel reprocessing plants.

In converting dose limits (e.g., 0.17 rem/yr) to radioactive release limits or concentration limits, it is necessary to treat each radioactive nuclide separately, and take into consideration how it moves within ecosystems. The concentration of a substance in a biological system can be quite different from that in the physical environment in which the biological system lives. Table 6.6 shows the factor by which concentrations of some radionuclides in fish, mollusks, and crustacea differ from the concentrations in the water in which the animals live. The factors vary from 1 (for tritium) to 10^4 (for ^{32}P). In some cases, substances become concentrated as they move up the food chain—they are selectively re-

Table 6.6 Concentration factors for various radionuclides in an aquatic ecosystem.

Isotope	Fish	Mollusks	Crustacea
^3H	1	1	1
^{32}P	10^4	10^4	10^4
^{60}Co	10^2	3×10^2	10^4
$^{131}I, ^{133}I$	20	10^2	10^2

Source: Adapted from *Engineering for Resolution of the Energy-Environment Dilemma*, ISBN 0-309-01943-5, Committee on Power Plant Siting, National Academy of Engineering, Washington, D.C., 1972, p. 184.

tained by whatever animal ate them, so that radionuclide concentrations that are "harmless" in the physical environment lead to increasingly dangerous concentrations at high trophic levels.

6.13   Dangers from Nuclear Energy

The fuel cycle for LWR's is shown in Fig. 6.4. In the early stages of the cycle (mining and milling), health hazards stem from the radioactivity of uranium. The air in the mines contains radioactive radon gas (from the uranium decay chain), and miners receive doses of radiation from inhaled radon and from radioactivity of the surrounding ore. Two kinds of cancer occur significantly more frequently in uranium miners than in the rest of the population. Uranium mining must be classified as a hazardous occupation, though it is less hazardous than deep-pit coal mining, both on a per-miner and on a per-unit-energy-extracted

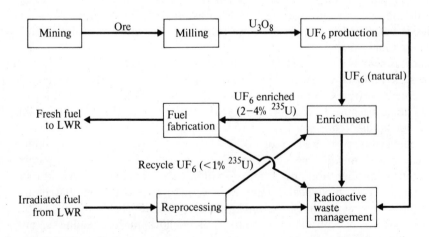

Fig. 6.4 Nuclear fuel cycle, light-water reactors. Uranium dioxide fueled, no plutonium recycle. (From *Environmental Survey of the Nuclear Fuel Cycle*, U.S. Atomic Energy Commission, Fuels and Materials, Directorate of Licensing, November 1972.)

basis. The waste material from mining and milling operations (overburden and processed ore) is mildly radioactive, and if discarded carelessly can blow into water supplies or cause other problems. In one instance, it was used as fill material on land for residential homes! The hazards in the mining and milling portion of the cycle represent an intensification of a natural hazard. With *reasonable precautions,* they need not be of major concern.

Once the uranium has been enriched, there exists the possibility of a *criticality accident.* That is, if too much enriched uranium becomes assembled in a small area, and if an appropriate moderator is present, the assembly could undergo a chain reaction and emit much radiation. (No explosion is possible because the assembly would blow itself apart before exploding.) The chances of a criticality accident are kept very small by the design of the containers and by the standard procedures used in handling enriched uranium.

The real dangers begin once the fuel enters the reactor, undergoes fission, and becomes highly radioactive. How much radioactivity is involved, for example, in a 1000-MW_e plant? Each fission delivers 200-MeV worth of energy and two fission fragments, each of which goes through a decay chain three to four steps long. To provide 3000 MW (necessary at 33 percent plant efficiency), there must be 10^{20} fissions/sec, implying 7×10^{20} decays/sec, or *2×10^{10} curies* in equilibrium operation.

That's a lot of radioactivity! Let's try to visualize it two different ways. (1) Comparison with nuclear weapons. Seven hours running of a 1000-MW_e plant produces energy equivalent to 20 kilotons of TNT, the same as a Hiroshima-sized bomb. It also produces as much radioactivity. (2) Potential for contamination. In a year's operation, a 1000 MW_e plant creates 3×10^6 curies of ^{90}Sr and 4×10^6 curies of ^{137}Cs. The legal limits for concentrations of these radionuclides in water are 300 picocuries/liter for ^{90}Sr and 2×10^4 pCi/liter for ^{137}Cs. Uniformly distributed, the ^{90}Sr could contaminate 10^{16} liters; the ^{137}Cs could contaminate 2×10^{14} liters. And 10^{16} liters is 2500 cubic miles; 2×10^{14} liters is 50 cubic miles. Lake Ontario has a volume of 400 cubic miles; the five Great Lakes have a combined volume of 6000 cubic miles. Thus a year's production of radioactivity from one reactor is more than enough to contaminate one of the Great Lakes! *It is clear that essentially all of the radioactivity produced in nuclear reactors must be contained.*

In subsequent sections, we will consider radioactive releases from a reactor during normal and abnormal operation, radioactive releases from a fuel reprocessing plant, storage of high-level radioactive wastes, and transportation accidents.

6.14   Radioactive Releases from Reactors

Normal Operation of LWR's

Radioactivity is contained within a reactor by several barriers. The uranium dioxide fuel pellets are enclosed in zirconium or stainless steel tubes called *clad-*

Fig. 6.5 Nuclear fuel assembly and shipping cannister. (Photo courtesy of Omaha Public Power District and the U.S. Atomic Energy Commission.)

ding. (A fuel assembly is shown in Fig. 6.5.) Most of the fission fragments remain within the fuel pellets, and most of those escaping from the pellets remain inside the cladding. Some, however, diffuse through the cladding or escape through pinhole leaks which may develop in the cladding. This allows a small fraction of the fission products to enter the primary coolant system. In addition, trace impurities in the cooling water capture neutrons, giving rise to further radioactivity. (While this second source is small compared to the total amount of fission products, it does not have the barriers of the fuel pellet and fuel-element cladding to penetrate.)

The primary cooling system is a closed system, and so provides a third barrier to escape of radioactivity. In a PWR, a leak from the primary cooling would be either to the containment building or to the secondary cooling system, meaning that there is yet a fourth barrier. In a BWR, radioactivity could leak from the primary cooling system to the environment through the condenser or the turbine. In both the PWR and BWR, the primary cooling water is taken through demineralizers and other mechanisms to remove the radioactivity. The removed radioactive nuclides are concentrated, solidified, and shipped "away" for storage. In this process, there are some low-level wastes which are diluted with cooling water and released to the environment. Liquid releases from a reactor range from 0.5 to 50 curies/yr depending upon the integrity of the cladding, among other things.

Gaseous radionuclides (principally isotopes of Kr, Xe, and I) that enter the primary cooling water are handled differently. In a BWR, the primary coolant is at low pressure during part of its cycle, so that a large fraction of the gases must be removed soon after they are produced. The iodine can be "caught" by various gas-cleaning schemes, but the noble gases are more difficult. Current practice is to filter out the iodine, hold up the gas for perhaps 30 minutes so the very short half-life nuclides can decay, and then release the gas to the environment. Annual releases are in the 100,000-curie range but they contain principally short half-life nuclides.

In a PWR, the gas can be kept in the primary coolant for a longer time, allowing more of the gas nuclei to decay. A smaller volume of gas must be processed, allowing a longer hold-up time (perhaps 40 days). Releases range from 10 to 10,000 curies/yr. The large range is due to the presence of some leaks that do not go to the holdup tanks. These vary from reactor to reactor. At the high end of the release range, releases are mostly short half-life nuclides (days), so they will not build up in the environment. At the low end of the release range, releases are mostly ^{85}Kr, which has an 11-year half-life.

PWR's have an additional source of radioactivity—tritium—for the following reason. As a reactor runs, its ^{235}U becomes depleted and fission products which capture neutrons accumulate. Both factors tend to reduce the reproduction constant k_e. This tendency must be compensated for, and the control rods have inadequate range for the task. In a BWR, this compensation is made by varying the bubble size of the boiling water, increasing moderator and therefore k_e with smaller bubbles. In a PWR, compensation is made by varying the boron content of the primary cooling-moderating water. When the fuel is fresh k_e would be high, so large amounts of boron are added to the water. Boron captures neutrons, lowering k_e. As the fuel becomes depleted and dirty, boron is removed from the water to compensate. Now, the neutron capture reaction on boron produces tritium, which is radioactive with a 12-year half-life. Tritium releases from a PWR range from 200 to 4000 curies/yr.

The radioactive releases from a light-water reactor under normal operation, along with figures for the radioactivity not released, are summarized in Table 6.7. The normal annual release of ^{137}Cs is in the range of 10^{-6} of annual production. For noble gases, the annual release is a larger fraction of annual production. For most other fission fragments it is a smaller fraction.

Under the 1971 AEC regulations (10CFR50, Appendix I), which LWR's can meet, normal radioactive releases do not pose a significant hazard.

Accidents at Reactors

All sorts of major and minor accidents are considered in the design of a reactor and in the licensing procedures for the reactor. The design incorporates con-

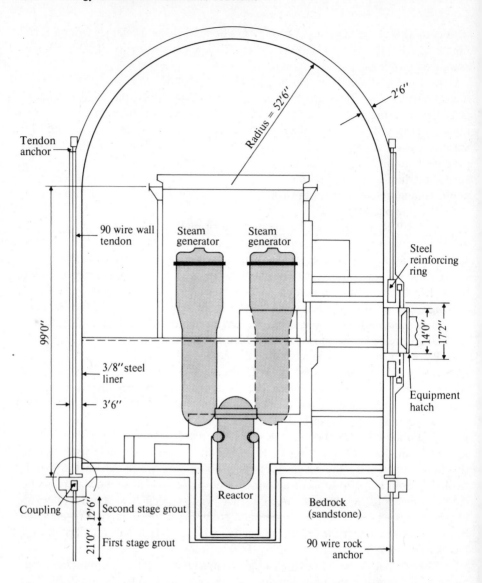

Fig. 6.6 Cross section of outer containment vessel. (From *Environmental Report,* R. E. Ginna Nuclear Power Plant, Rochester Gas and Electric Corporation, 1972.)

siderable redundancy in the safety features—there are parallel monitoring systems, auxiliary sources of electricity for the control equipment, and so on. For each accident considered, failure of several independent components is required before the consequences of the accident become serious. As last-ditch protection, the reactor is surrounded by an outer containment vessel, a thick hemispherical

Table 6.7   Radioactive releases, radionuclide production, and radioactivity of an LWR under normal conditions.

Annual releases	Liquid (exclusive of tritium)	0.5–50 curies (^{137}Cs, others)
	Tritium (PWR)	200–4000 curies
	Gaseous (PWR)	10–10,000 curies (noble gases)
	Gaseous (BWR)	50,000–500,000 curies (noble gases, short half-life)
	Annual production of ^{137}Cs	4×10^6 curies
	Content of reactor one hour after shutdown	6×10^9 curies

Source: Adapted from *Engineering for Resolution of the Energy-Environment Dilemma,* ISBN 0-309-01943-5, Committee on Power Plant Siting, National Academy of Engineering, Washington, D.C., 1972.

concrete shell two to three feet thick and about 100 feet in diameter (see Figs. 6.6 and 6.7). The containment is capable of withstanding considerable external abuse (tornados, mild earthquakes), and also internal abuse (release of primary cooling water as steam from reactor to containment vessel). Licensing procedures require that the overall design of the plant be such that the "maximum credible accident" would not result in serious radiation exposures to nearby populations.

All this is fine, if only the anticipated accidents occur and if they proceed according to the plans set for them. But it is in the nature of accidents that they are unexpected and often caused by incorrect thinking about some process or by neglect of some important point.

Fig. 6.7 Dresden Nuclear Power Station, showing outer containment vessel around an 800-MW$_e$ BWR. (Photo courtesy of Commonwealth Edison Company and the U.S. Atomic Energy Commission.)

Currently, a type of accident causing much concern[10] is the *loss of cooling accident* (LOCA). Suppose there is a large break in a large pipe in the primary coolant system. At high pressure, the water in the reactor vessel will rush out. As the pressure drops, water will flash to steam, continuing the ejection of water. Soon the water level will drop below the fuel elements and they will be without coolant. Since there is also no moderator, the chain reaction stops. But heat is still generated in the fuel by radioactive decay—8 percent of full power when the chain reaction stops, and 1 percent of full power a few hours later. The temperature of the fuel elements rises rapidly; cooling must be restored within a minute or so or the fuel elements will melt. There is an *emergency core cooling system* (ECCS) with a lot of redundancy. However, some people have questioned whether ECCS will work. Steam from the high-temperature fuel may drive the emergency cooling water away. Fuel elements damaged by rising temperatures may obstruct water passages. To date, tests of ECCS have not been convincing. If ECCS does not work, the cladding will melt; the fuel may melt, collect in a puddle in the bottom of the reactor, maintain its heat from radioactivity, and burn and melt its way through the reactor vessel and on down into the earth (*China syndrome*). In this case, the radioactivity would no longer be contained. Gaseous decay products could enter the atmosphere, ground water could be contaminated, and so on. The LOCA is clearly a very serious accident.

The probability of a serious accident is claimed to be very low. This is based in part on the assumption that, since multiple failures are required for a serious accident, and since the probability of an individual failure is small, the probability for the serious accident will be the product of the small individual probabilities, therefore very small. This thinking is correct only if the individual failures are totally independent. It is far more likely that the multiple failures would result from a common cause. For example, a severe electrical storm could damage several components at the same time. Two parallel systems (such as two diesel-driven auxiliary generators for emergency power) could both fail to function due to an identical mistake in maintenance by a single individual. A fire could damage several "independent" control devices simultaneously. A human error by a reactor operator could be followed by another human error by the same, now excited, operator. Finally, the probability of a *deliberate* accident—sabotage, vandalism, terrorism—may well be considerably higher than any of the foregoing possibilities. It is likely that if a serious reactor accident does occur, it will not be because a highly improbable event has happened, but rather because someone's estimate of probabilities was in error. (The "maximum credible accident" for the Fermi breeder reactor was the melting of one fuel assembly; an accident that actually occurred there resulted in the melting of three assemblies. In hindsight, this "incredible" accident was not the result of highly improbable happenings. During construction, a baffle plate had been installed with bolts rather than being welded in place as the design specified. The bolts failed and the plate blocked sodium flow.

Because the large amount of radioactive material in a reactor provides the *potential* for a very serious accident, and because there are uncertainties associated with predicting the occurrence of accidents, the subject of reactor accidents will (and should) continue to be of concern.

6.15   Fuel Reprocessing, Waste Storage, and Transportation

Fuel Reprocessing

After three to four years in a reactor, the ^{235}U fuel becomes depleted and becomes contaminated with fission fragments, and must be reprocessed. The procedure is the following. The reactor is shut down and time is allowed for the short half-life radioactivity to die away. The old fuel elements are removed and replaced. (Typically, one-third to one-quarter of the fuel elements are replaced, once a year.) The old fuel elements are stored at the reactor site under water (for cooling) for about three months. Then they are transferred to the fuel reprocessing plant and stored there a while longer. Five months after removal from the reactor, one year's used fuel from a 1000-MW$_e$ plant contains 1.5×10^8 curies. The fuel elements are chopped up, dissolved in nitric acid, and manipulated chemically. The aim is to salvage uranium and plutonium, to release ^{85}Kr and tritium to the environment, and to store everything else *in perpetuity*. Typical releases for one year's fuel from a 1000-MW$_e$ plant are 2×10^4 curies of tritium (12.3-year half-life), 0.35×10^6 curies of ^{85}Kr (10.7-year half-life), and 1 to 5 curies of miscellaneous other substances. Note that the releases per reactor in the fuel reprocessing step are considerably larger than at the reactor. Also, these are longer half-life nuclides, so the radioactivity will build up as more fuel is reprocessed. If nuclear power grows as predicted, and if ^{85}Kr continues to be released into the atmosphere, in 100 years ^{85}Kr radioactivity will constitute a radiation level comparable to natural background. Clearly, long before then both ^{85}Kr and tritium must be stored rather than released. Cryogenic techniques should allow capture of the ^{85}Kr. Isolating tritium appears more difficult.

The possibility for accidents at the fuel reprocessing plant exists. While the amount of radioactivity is high, it is less than in a reactor, and the power level and operating pressure are much lower. However, the number of barriers between the radioactive material and the environment are fewer, and the radioactivity has a longer half-life, capable therefore of causing more long-term harm.

Waste Storage

Low-level radioactive solid wastes are buried in near-surface trenches (20 ft deep) at special sites where the topography, meteorology, and hydrology are such that migration of the radioactivity is not anticipated. The intention is that the land at a radioactive-waste disposal site is committed "forever." The land requirements for the low-level waste from a 1000-MW$_e$ plant and the fuel cycle activities attributable to that plant are about 0.2 acres per year. While the procedure just described is certainly sloppy, and isolation of the radioactivity from the environment would be preferable, it is unlikely that current procedures will lead to significant problems.

High-level radioactive wastes—the wastes recovered in the fuel reprocessing step—are quite a different matter. Up to this point we have been talking about the small fraction of radioactivity that leaked out. Now we must face the large fraction—essentially 100 percent—that has been successfully contained. As mentioned previously, this is 1.5×10^8 curies per 1000 MW_e-year, after fuel reprocessing. It must be isolated from the environment until it decays to insignificant levels. Ten half-lives will reduce it by a factor of a thousand—not enough. Twenty half-lives will reduce it by a factor of a million—more what's needed. With the 30-year half-lives of ^{90}Sr and ^{137}Cs, we are talking about 600 years. Taking on a 600-year commitment is not something that should be entered into lightly.

At present, high-level wastes are stored as liquids in stainless-steel tanks. The tanks must be cooled; otherwise the heat from radioactivity would cause the liquid to boil. Tanks have an expected life of about twenty years; if a leak develops, the liquid must be transferred to another tank. Regulations as of 1971 call for converting the high-level wastes to solid form within five years after fuel reprocessing, and shipment to a federal repository within ten years. Said federal repository does not yet exist. Plans call for something by 1980.

The most promising scheme for perpetual storage is storage in bedded salt formations deep underground.[11] Under natural conditions these formations are dry and well separated from flowing water, and they are often located in areas of low seismic activity. Salt is plastic, so fractures should close; it has good compressive strength, so caverns can be mined in it at considerable depths. The intention would be to place the solidified wastes in small chambers mined from the salt, then fill the chamber with loose salt. Heat from radioactivity would consolidate the salt, thereby permanently isolating the wastes (see Fig. 6.8). However, the integrity of many salt formations has been violated by human activity (mining and exploratory drilling), so that isolation from water is not certain. A site in Lyons, Kansas, tentatively selected for a repository, had to be discarded when water due to previous mining activity was discovered nearby.

The need to find a perpetual storage scheme that requires a minimum of surveillance is considerable. It is not clear that human political institutions can do *anything* for a period as long as 600 years. It must be realized that any scheme will require *some* surveillance. In developing nuclear energy, we are burdening future generations for centuries into the future with our high-level wastes. They may dislike us for that!

Transportation

High-level wastes move from reactor to fuel reprocessing plant to permanent repository by way of the highways and railroads. They are packaged in special containers to withstand accidents, but it is not clear that a serious accident wouldn't break the container, releasing radioactivity to the environment.

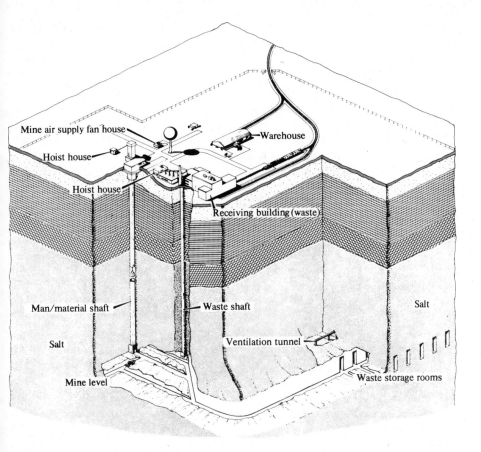

Fig. 6.8 Artist's conception of a bedded salt repository for solidified high-level radioactive wastes. The storage area is 1000 to 2000 feet underground. (Photo courtesy of Oak Ridge National Laboratory and the U.S. Atomic Energy Commission.)

6.16   Breeder Reactors

While the problems discussed in the preceding sections apply in a general way to all fission reactors, we have framed our discussion in terms of LWR's, since they are what is currently in use, and also what is coming into use in a big way. Present U.S. energy policy calls for the use of LMFBR's beginning around 1985, and a sizable research and development effort is devoted to that end. It is appropriate to consider at this point in time what environmental problems these new reactors may cause. We will focus on the differences between the LMFBR's and the LWR's.

Plutonium-239 is a radiological poison. It decays by emission of a 5-MeV α-particle, with a half-life of 24,000 years. If ingested, plutonium is absorbed by bone marrow. The short range and high specific ionization of α-particles results in a high, localized dose. Extreme care must be taken in fuel-element fabrication and in fuel reprocessing to keep plutonium completely contained.

Plutonium-239 is the material from which nuclear bombs are made. If our energy supply becomes based on LMFBR's, there will be a large amount of ^{239}Pu in circulation and the opportunities for theft will be numerous. Likely someone who shouldn't will try to make a bomb.

The fuel elements for LMFBR's are highly enriched by comparison to those for LWR's. As a result, a criticality accident at some stage in the fuel cycle is no longer so unlikely.

Liquid sodium, which is used as coolant, is highly reactive both with water and with air. Should a small leak develop between the reactor vessel or the sodium plumbing and the atmosphere, this reactivity could enlarge the leak, and also would make repair difficult.

As discussed in Section 5.2, LMFBR's lack the inherent stability of LWR's. As a result, there is a greater chance for a reactor malfunction to escalate into a more serious problem. For example, there is concern that a void in the liquid sodium caused by blockage of sodium flow could cause a portion of the reactor to "run away," overheat, damage fuel elements, and lead to a very serious accident.

On the positive side, LMFBR's would operate at much lower pressures than do LWR's, and this factor reduces the likelihood of leaks and ruptures. LMFBR's also would operate at higher temperatures, and as a result are expected to have efficiencies for converting heat to electricity of 40 percent (comparable to modern fossil-fuel plants) rather than the 33 percent of LWR's. The higher efficiency means less thermal pollution per unit of electricity produced. Finally and obviously, breeder reactors would conserve natural resources. By using all the uranium rather than just ^{235}U, the amount of uranium needed is much reduced.

6.17   Summary of Nuclear Energy Hazards

Summarizing the discussion of Sections 6.13 through 6.15:

1. Mining and milling of uranium entail hazards that are intensifications of natural hazards. With reasonable precautions, they need not be of major concern.

2. With enrichments used in LWR's, careful procedures can keep the probability of a criticality accident negligibly small.

3. Radioactive releases from an LWR under normal conditions can be and are being kept quite low, and so are not a problem.

4. In spite of all precautions, accidents at reactors are always a possibility. Results of a major accident could be serious.

5. The radioactive releases from a fuel reprocessing plant under normal conditions are too high. For the long haul, both ^{85}Kr and tritium will have to be contained.

6. Long-term storage of high-level radioactive wastes is an unsolved problem, and a very serious one.

7. With increasing volumes of high-level radioactive materials moving about the country, a transportation accident involving the release of radioactivity to the environment appears likely to occur sooner or later.

The above conclusions refer to light-water reactors as currently in use for electric power production in the United States. For a different type of reactor— say the fast breeder—some of the conclusions are altered (see Section 6.16).

Nuclear energy is potentially extremely hazardous. Its utilization requires extreme care in all phases of operation; such care is exemplified by radioactive releases being kept down to the order of 10^{-6} of produced radioactive material. The nuclear energy industry's good record to date is due in large part to an unusually high level of supervision by the Atomic Energy Commission (AEC), which boasted an *unprecedented* record of enforcement. This level of supervision and enforcement must be maintained. (If it is not, I will have to retract my optimistic conclusion about radioactive releases at power plants (3), and will have to express considerably more concern about accidents (4), fuel reprocessing (5), storage (6), and transportation (7).)

In the past, regulation has always *followed* new technology. The sequence has been (a) new technology, (b) increased scale of use, (c) disaster, and (d) regulation. Nuclear energy is unique in that it is the only case where a new technology was matched *in advance* by new methods of regulation. Credit for this achievement must be shared by the AEC, the Joint Committee on Atomic Energy (U.S. Congress), and the many individual scientists, organizations of scientists, and members of the lay public who have exerted pressure on the AEC to do its job well. With the federal energy reorganization (see Section 6.4), the NRC assumed the regulatory functions of the AEC. NRC bears close watching to ensure that its performance does not fall below that of the AEC.

The hazards discussed have all been in the human health and safety category. Damage to ecosystems has not been discussed. In the case of radioactivity, it turns out that within the framework of current U.S. planning, ensuring human health and safety will also ensure negligible damage to ecosystems. This is so because (1) individual human beings are not significantly less sensitive to radiation than are individual members of most other species, (2) a level of health and safety damage far in excess of what is acceptable for human beings is required to damage an ecosystem (because with people, we worry about individuals, but with all other species, we worry only about the well-being of the species as a whole), and (3) the U.S. radiation management does not subject other species to radiation levels significantly above those received by the occasionally exposed individual of the general public. One can imagine practices contrary to point (3), for example, dumping high-level radioactive wastes into the deep ocean. Such practices would require a careful consideration of ecosystem damage in addition to human health and safety considerations.

PART D / AIR POLLUTION

Air pollution is the classic example of, "The solution to pollution is dilution." To a very large extent, air pollution is an urban problem—"If the stuff could just be spread around a bit, it would be harmless." However, as man's use of energy grows, the scale of air pollution grows, and at some point ceases to be only a local problem. Following a general survey of the field, we will examine two pollutants, sulfur dioxide and carbon dioxide, that are rapidly becoming global problems. I have chosen to concentrate on regional/global problems rather than local problems not because they are more serious (presently the opposite is true), but rather because most other discussions of air pollution focus on the local problems. Particulate air pollution is another global problem worthy of consideration. It is omitted here because I have nothing useful to say about it—it does not lend itself to a brief, general treatment.

6.18   Air Pollution: A Short Course

What Are They?

Primary air pollutants include:
a) Gases—principally CO, CO_2, assorted hydrocarbons, H_2S, SO_2, various oxides of nitrogen, and
b) Particulate matter—dusts, mists, soot, lead, and all sorts of industrial substances (e.g., sulfates, nitrates, fluoride).

Once in the air, the primary pollutants take part in a wide range of chemical reactions, forming new products (secondary pollutants). Often these secondary pollutants cause greater damage than is caused directly by the primary pollutants. Some examples of secondary pollutants include:

SO_2 ($2H_2S + 3O_2 \rightarrow 2SO_2 + 2H_2O$),

Sulfuric acid mist ($2SO_2 + O_2 \rightarrow 2SO_3$; $SO_3 + H_2O \rightarrow H_2SO_4$),

Ozone ($NO_2 + \text{sunlight} \rightarrow NO + O$; $O + O_2 \rightarrow O_3$), and

Smog (hydrocarbons + NO_2 + sunlight $\xrightarrow{\text{many intermediate steps}}$ peroxyacyl nitrates + formaldehyde + other).

In assessing the hazards of a primary pollutant, one must take into account the secondary pollutants derived from it. This, of course, depends on what other components, pollutants and otherwise, are present in the atmosphere. For example, one ill effect of hydrocarbons (smog formation) depends on the presence of NO_2 (another pollutant) and also on the presence of sunlight (usually considered beneficial).

Where Do They Come From?

Sources of air pollutants are listed in Table 6.8; a more detailed breakdown is given for sulfur dioxide in Table 6.9. The dominance of energy utilization (transportation and combustion in stationary sources) is apparent.

The physical processes involved in the production of CO and CO_2 are the incomplete and complete combustion of the carbon content of fossil fuel, respectively. The hydrocarbons come from evaporation of hydrocarbon fuels and from

Table 6.8   Estimated U.S. nationwide air-polluting emissions, 1968 (in millions of tons per year).

Source	Carbon monoxide	Particulates	Sulfur oxides	Hydrocarbons	Nitrogen oxides	Total
Transportation	63.8	1.2	0.8	16.6	8.1	90.5
Fuel combustion in stationary sources	1.9	8.9	24.4	.7	10.0	45.9
Industrial processes	9.7	7.5	7.3	4.6	.2	29.3
Solid-waste disposal	7.8	1.1	.1	1.6	.6	11.2
Miscellaneous[a]	16.9	9.6	.6	8.5	1.7	37.3
Total	100.1	28.3	33.2	32.0	20.6	214.2

[a] Primarily forest fires, agricultural burning, coal waste fires.

Source: NAPCA Inventory of Air Pollutant Emission (1970), quoted in The First Annual Report of the Council on Environmental Quality, p. 63.

Table 6.9   Sulfur dioxide emitted over the United States, Great Britain, and West Germany (in millions of tons per year).

Source	U.S., 1966	Great Britain,[a] 1945	West Germany, 1962
Power generation—coal	11.9	1.1	1.4
Power generation—oil	1.2	—	—
Other coal combustion	4.7	3.3	—
Other oil combustion	4.4	0.5	—
Smelting of ores	3.5	—	—
Oil refining operation	1.5	—	—
Coke processing	0.5	0.1	—
Sulfuric acid manufacturing	0.5	—	—
Coal refuse banks	0.1	—	—
Refuse incineration	0.1	—	—
Miscellaneous sources[b]	0.1	0.1	—
	28.6	5.1	—

[a] Total for 1963 was 3.4.
[b] Natural gas, LPG, pulp and paper mills, chemical manufacture, etc.

Source: Arthur C. Stern (ed.), *Air Pollution*, Vol. III. New York: Academic Press, 1968, p. 8. By permission.

engine blowby. Sulfur dioxide comes from combustion of the sulfur content of fossil fuel. (Sulfur is one of the elements present in living matter, and is retained when this matter is fossilized into coal or oil. See Section 6.19.) Oxides of nitrogen are formed whenever N_2 and O_2 are both present at high temperatures. (The reaction $N_2 + O_2 \rightleftarrows 2NO$ has an equilibrium far to the left at the atmosphere's ambient temperature. At elevated temperatures, the equilibrium shifts toward the right, so nitric oxide is formed from air in high-temperature combustion processes. Further, at lower temperatures the approach to equilibrium is very slow, so if the

NO is formed at high temperatures and subsequently rapidly cooled, oxides of nitrogen will persist for some time. In the atmosphere NO is oxidized to NO_2).

Particulates enter the atmosphere during the combustion of fossil fuel as the noncombustible material (ash) is carried upward along with combustion gases. They also appear as secondary pollutants—nitrates and sulfates from NO_2 and SO_2.

The preceding discussion and tables refer to man-made sources of air pollution. We must also consider natural sources of the same substances. On a global scale:

a) Carbon monoxide is essentially all due to human activities;

b) Carbon dioxide is predominantly natural, but man-made CO_2 is significant (see Section 6.20).

c) About 15 percent of the hydrocarbons entering the atmosphere are the result of human activities. Natural sources are forest vegetation, which gives off turpenes, and bacterial decomposition of organic matter, which gives off methane.

d) Perhaps 25 percent of the SO_2 in the atmosphere comes from human activities (see Section 6.19). The dominant natural source is the decay of organic matter, giving rise to H_2S which is oxidized to SO_2 in the atmosphere.

e) For oxides of nitrogen, the situation is unclear. Man-made sources dominate locally but are probably unimportant globally.

f) Man-made sources of particulates contribute about 20 percent of the global total. Natural sources include sea spray, volcanoes, windblown dust, and the conversion of naturally occurring gases into particles (e.g., SO_2 to sulfate particles and sulfuric acid mist).

Man-made sources of air pollutants are heavily concentrated in urban areas. If pollutants don't disperse rapidly, then the pollutant concentrations there will be well above average. As a result, man-made air pollutants greatly exceed natural pollutants in urban areas.

How Do They Behave and Move in the Atmosphere? Where Do They Go?

The bulk motion of pollutants follows the bulk motion of air, i.e., wind and weather. With a good wind, pollutants will be carried away from their source and dispersed over a large area. If the air is still, as occurs during a thermal inversion, pollutants are not carried away and can build up to dangerous levels.

Particulates settle out of the atmosphere under the force of gravity and are carried out by rain and snow. The larger particles are removed most rapidly. The settling rate under gravity is governed by Stoke's Law, which says that the terminal velocity of a spherical particle is proportional to the square of the particle diameter. A 1-micron diameter particle has a terminal velocity of about 0.01 cm/ sec or 0.3 m/hr. Those particles of diameter much above 10μ are quickly removed by gravity, whereas those smaller than 1μ are not removed by gravity to any appreciable extent. Sometimes small particles coalesce to form larger particles which then settle out. (Pulverized coal fly-ash ranges from 1μ to 50μ in diameter; sulfur dioxide mist ranges from 0.3μ to 3μ in diameter.)

Gaseous pollutants react in various ways, often becoming particulates. For example, sulfur dioxide is converted to sulfuric acid mist or sulfates. The sulfur spends a total of one to two weeks in the atmosphere before being removed. On the other hand, carbon monoxide has a residence time in the atmosphere of a few years. It is suspected (though not established) that carbon monoxide is converted to carbon dioxide.

What Harm Do They Cause?

On a global scale, carbon dioxide and particulates can cause climate alterations. Particulates increase the earth's albedo, and thus lower the temperature. Increases in carbon dioxide lead to an enhanced greenhouse effect, thus raising the temperature. Present levels of particulates and CO_2 due to human activities might each result in a change in the average temperature on the order of 0.2°C, and the two effects tend to cancel, resulting in an even smaller change. At present levels, the effects are too small to observe in the presence of natural changes, and are sufficiently small as to cause no harm. Increases in CO_2 as projected for the not-too-distant future would cause major climate alterations (see Section 6.20).

On a regional scale, rainfall over much of western Europe is acidic (ph 4 to 5) due to SO_2-generated sulfuric acid. It is endangering forests and aquatic life (see Section 6.19).

To date, most of the damage has occurred at the local level in individual industrial and urban areas. It includes the following.

a) Acute health effects from high-level pollution. This has occurred during extended inversions. The three-day episode in Donora, Pennsylvania (1948) resulted in 600 illnesses and 20 deaths in a population of 14,000. One episode in London in 1952 resulted in 3000 to 4000 deaths.
b) Chronic disease (bronchitis, emphysema, and shortened lifespan) from long-term low-level exposures.
c) Eye irritation, offensive odors, general discomfort to people.
d) Damage to vegetation—crops and trees.
e) Injury to livestock. (Fluorides cause fluorosis in cattle.)
f) Deterioration of materials (buildings, paints, and bridge structures).
g) Soiling of materials (clothing, curtains, etc.).
h) Reduction in visibility (both a safety matter and an aesthetic one).

The EPA has estimated some of the economic costs of air pollution[12] and finds that (1) the health costs (medical bills, work loss) are 6 billion dollars annually, (2) damage to vegetation and materials is 5 billion dollars annually, and (3) reduced property values are 5 billion dollars annually. It must be stressed that estimates of this sort are very difficult to make, and so these figures are rough. It must also be stressed that not all the costs are included. Health costs do not include those for discomfort, anxiety, or suffering. Other psychological costs (items (c) and (h) above) appear only as they are reflected in reduced property values.

What Can Be Done About It?

Dilution as an abatement technique made considerable sense when the affected region was the immediate neighborhood of a power plant or factory. The tall smokestack, which allows some dispersal of the effluent, was the standard approach. With the regions in trouble being entire metropolitan areas, and the sources of pollution being several power plants plus all the automobiles in the area, dilution by tall stacks is no longer adequate. A modern variation on that theme is to locate the power plant well away from the metropolitan area being served, and bring the electricity in on transmission lines. The most infamous example is the Four Corners plant which serves Los Angeles, but is located near the meeting point of New Mexico, Arizona, Colorado, and Utah in a sparsely settled desert area. By conventional cost-benefit analysis, this appears to be a proper site. However, the costs accrue to one group of people and the benefits to another, which seems unjust. Further, the assumption that polluted air is more acceptable in a low population-density area than a high population-density area is debatable. Health problems and materials damage are less, but damage to natural ecosystems is probably greater. (Those in an urban area are usually already lowered in quality.) Also, psychological/aesthetic losses may be larger. (Exceptionally clean air is one of the major benefits of low population-density areas in general and the southwest United States in particular. Somewhat dirtier air can be tolerated in the cities.) Independent of one's views of the "Four Corners approach," dilution is not by itself a sufficient abatement technique.

If the pollutant is part of the fuel (as in fly-ash and sulfur), one can consider (a) using fuels with lower pollutant content, (b) removing the pollutant from the fuel, or (c) removing the pollutant during the combustion process or from the gases going up the smokestack.

Concerning the first option, natural gas is essentially free of both ash and sulfur; oil is very low in ash and contains variable amounts of sulfur (0.5 percent to 4 percent), depending on its origin. Coal contains variable but substantial amounts of ash and variable amounts of sulfur (0.5 percent to 4 percent), depending on its origin. A currently practiced SO_2 abatement technique is the use of low-sulfur fuels. This approach limits the usable supply of fuel and increases its dollar cost.

Concerning the second option, in the process of refining, the sulfur content of crude oil is removed from the lighter distillates (gasoline and home heating oil) and is concentrated in the residual oil. The technology exists today for removing sulfur from residual oil, and some refineries are equipped to do it. The only shortcomings to this approach are refinery capacity and dollar cost. The technologies for removing sulfur and ash from coal (e.g., by gasifying or liquefication) are approaching the pilot plant stage, but are not yet in commercial use.

Concerning the third option, stack-gas cleanup is the standard approach for removing fly-ash from coal. Filters, electrostatic precipitators, and other such devices have been removing increasingly higher percentages, so that now 99.5 percent fly-ash removal (by weight) can be achieved. Quoting percentage removal by weight can be misleading, however, since the larger particles are more efficiently removed. These are also more quickly removed from the atmosphere by natural processes. It is precisely those particles that persist in the atmosphere for a long time that are least efficiently removed by the filters, etc. Progress in removing sulfur from the stack gas or during the combustion process has been quite disappointing. Despite high hopes a few years ago, there are still no reliable, commercially proven procedures.

A typical power plant uses more than one abatement technique. Tall stacks and electrostatic precipitators are standard. Use of low-sulfur fuel is common. A large power plant is shown in Fig. 6.9.

For pollutants that are a result of a "defect" in the combusion process (NO, CO, hydrocarbons), one can either modify the combustion process or use "tricks" (e.g., catalysts, afterburners) on the post-combustion gases. Both

Fig. 6.9 Cumberland Steam-Electric Power Plant, the largest coal-fired power plant in the TVA system. There are two units, each with a capacity of 1300 MW$_e$. Note the tall stacks for air pollution dispersal and the large area for coal storage. Coal is delivered by barge (right foreground), unloaded, and moved by a system of conveyors. (Photo courtesy of the Tennessee Valley Authority.)

approaches are now being tried on automobile engines, the chief offender in this category.

Carbon dioxide is a principal product of complete combustion of fossil fuel. It cannot be thought of as an impurity (as with sulfur and ash) or a defect in the combustion process. If fossil fuels are used, CO_2 will be produced in large quantities. It is difficult to conceive of an abatement technique.

Air pollution differs from radioactivity in that abatement schemes aimed at protecting human health and safety will not necessarily protect natural ecosystems or psychological/aesthetic values. This is so because (1) many species of plants are considerably more sensitive to air pollution than humans are, (2) air pollution causes direct aesthetic damage whereas radioactivity does not, and (3) the abatement technique of discharging the pollutant in natural areas with sparse human populations is contemplated for air pollution but not for radioactivity.

6.19   Sulfur Dioxide—A Regional Problem

There is no general acceptance of the large-scale nature of the SO_2 problem. Many people agree with Hottel and Howard,[13] that "... combustion generated SO_2 may present a dispersion problem or a local pollution problem, but it is not a global pollution problem; the life of SO_2 is too short." In assessing the scale of the SO_2 problem, we begin by comparing the magnitude of the man-made and natural sources of atmospheric SO_2. Direct natural inputs of SO_2 into the atmosphere are believed to be very small. Most atmospheric SO_2 of natural origin results from H_2S entering the atmosphere during decay of organic matter; H_2S in the atmosphere is rapidly oxidized to SO_2. Direct estimates of natural H_2S entering the atmosphere are rather uncertain, so we use an indirect approach, a modification of one due to Hottel and Howard.[14]

Hottel and Howard obtain their estimate of the ratio of man-produced to natural atmospheric sulfur by assuming that the ratio is equal to the ratio of man-produced CO_2 to CO_2 introduced naturally from organic decay. This is equivalent to assuming that the sulfur-to-carbon ratio of the average fuel is the same as the sulfur-to-carbon ratio of organic matter. Since fossil fuels come from organic matter, that is a reasonable first guess. However, it turns out to be off by a factor of 5; the average fuel has a sulfur-to-carbon ratio of ~0.01, by weight, while organic matter has a ratio of 0.002.[15] Therefore, Hottel and Howard's approach underestimates the man-produced-to-natural-atmospheric-sulfur ratio by a factor of 5, and should be corrected accordingly.

The magnitude of man-produced CO_2 is 4×10^9 metric tons of carbon per year (1970). The magnitude of CO_2 from organic decay is less well known. Estimates vary from 22 to 151×10^9 tons of carbon per year, with a "best guess" at 90×10^9. (Note that carbon from organic decay is equal to carbon fixed in photosynthesis. These estimates correspond to worldwide net photosynthesis of 28 to 200×10^{12} watts, compared to the value of 40×10^{12} watts shown in Fig. 4.1.) Therefore the ratio of man-made to natural atmospheric sulfur is in the range of 1:8 to 1:1, with the "best guess" being 1:4. Estimates by others, certainly not independent of the considerations given above, are in the same range.[16] By anyone's estimate, human contributions to atmospheric sulfur exceed the "one-percent law" by a very considerable amount.

In the atmosphere, SO_2 is oxidized to SO_3, which reacts with water to form sulfuric acid mist. This in turn may react with other material to form sulfate particles. The average residence time for sulfur in the atmosphere is short—in the neighborhood of ten days.

From the residence time, the size of the region over which sulfur from a point source will disperse can be estimated. Assuming sulfur is transported with the same speed as the weather (about 20 mph average), it may travel as much as 5000 miles downwind in the ten days. If the sulfur is injected into the atmosphere close to ground level, it may move much slower (4 mph is a reasonable average wind speed) and travel perhaps as much as 1000 miles downwind. Sulfur deposition will be greatest close to the point source, and will decrease downwind (e.g., roughly exponentially) with a characteristic distance of 1000 to 5000 miles. It therefore affects a large region but is not uniformly distributed around the globe. Since the prevailing winds travel west-to-east, the distribution will be greatest in this direction and much more restricted in the north-south directions.

Since energy use by man is unevenly distributed over the globe, SO_2 distribution will also be uneven. By far the most SO_2 is released in the northern hemisphere between 30 and 60 degrees latitude. More specifically, energy use per unit area in the eastern United States and in central Western Europe is about twenty times larger than the world average for land surfaces. For these regions and several hundred miles downwind, man-produced sulfur exceeds natural sulfur by a factor of perhaps 5, and observable effects from excess sulfur deposition are to be expected.

Marked effects have been noticed in Europe. As described by Erikson[17] the rainfall there has become increasingly acidic, and lakes in Sweden have become acidic to the point of affecting aquatic life. Acidic rainfall is harmful to soil also. It is estimated that at present levels of sulfuric acid deposition, some of the forest soils will be exhausted of nutrients in 100 years; present rates of energy growth would shorten this time to 40 years.

Technically, the damage just described is not caused by air pollution *per se.* Air is merely the transport medium—the trouble arises when the sulfur accumulates in water and on land. It appears that if increased energy use is not compensated for by sulfur removal techniques, large natural regions will be adversely affected over the next few decades.

6.20   Carbon Dioxide—A Global Problem

Utilizing fossil fuels as an energy source creates carbon dioxide. This is not an impurity, an incidental, or an accident, as SO_2 or CO might be considered to be. It is basic and unavoidable.

Most pollutants first cause trouble on a local scale, and only after they become very serious local problems do they have any global effects. Carbon dioxide, on the other hand, causes no local problems and is often not considered a pollutant. If it causes any problems at all, it will do so on a global scale.

The circulation of carbon in the biosphere is shown in Fig. 6.10. Both the amounts of carbon in any given place and the annual flows are shown. Some of these numbers are not firmly established, but the picture is qualitatively correct. Two features should be noted. (a) The carbon flow from fossil-fuel combustion is comparable to some of the natural flows and is not negligibly small compared to their sum. It represents about 4 percent of the CO_2 entering the atmosphere. (b) The amount of carbon stored as recoverable fossil fuel is large compared to most other storages. The two exceptions—deep ocean and sediment—are weakly coupled to the rest of the system. From the preceding, we can anticipate that the combustion of fossil fuels will cause an increase in the carbon content of one or more of (1) the atmosphere, (2) the sea surface layers, and (3) the land biomass.

The concentration of CO_2 in the atmosphere has been increasing. Measurements and extrapolations suggest that the pre-fossil-fuel-era value was 290 ppm (by volume), the 1970 value was 320 ppm, and the average rate of increase between 1960 and 1970 was 0.7 ppm/yr. Had all the CO_2 from fossil-fuel consumption between 1960 and 1970 remained in the atmosphere, the increase during that time period would have been twice as large. About half the carbon added by fossil-fuel combustion apparently goes to the ocean or to the land biomass. The ratio of distribution between these two is not well understood. A simple model,[18] which agrees with existing data and assumes fossil-fuel use will continue to grow at 4 percent per year, leads to the conclusion that CO_2 concentration will reach 375 ppm by the year 2000.

No major adverse effects are anticipated as a result of an increase in the carbon content of the land biomass or the ocean. The adverse effect anticipated as a result of an increase in the concentration of atmospheric CO_2 is climate modification due to an enhanced greenhouse effect (see Sections 2.5 and 2.7). Calculations by Manabe,[19] which include the positive feedback effect of increased water vapor with higher temperatures, yield a $0.5°C$ surface warming for a change from 320 ppm to 375 ppm and a $2°C$ warming for a doubling of atmospheric CO_2. The former temperature increase is significant and the latter is a very large change, which undoubtedly would cause major global climate changes, among them the melting of polar ice.

If we limit our fossil-fuel use to that which will double atmospheric CO_2—already a rash thing to do—then assuming half of the fossil-fuel carbon stays in the atmosphere, we can burn only 1400 billion more tons of fossil-fuel carbon, which is about 20 percent of the world's coal supplies (see Section 4.4). At present levels of use that would take 300 years; at present growth rates, 65 years.

The preceding limit assumes that fossil fuels are burned "suddenly," over a period of a few hundred years. If they were burned "slowly," then the CO_2 concentration in the deep ocean would have time to come to equilibrium with that

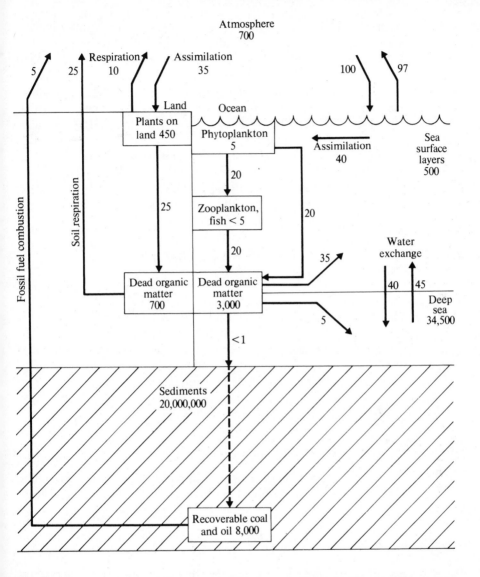

Fig. 6.10 Circulation of carbon in the biosphere. Quantities are either in billions of tons of carbon (for "standing crops") or in billions of tons of carbon per year (for flows). (From Bert Bolin, "The Carbon Cycle," *Sci. Amer.* **223**, p. 130. Copyright © 1970 by Scientific American, Inc. All rights reserved.)

of the surface layers, and a much larger total amount of fuel could be burned for the same atmospheric CO_2 increase. The time required for this equilibrium is of the order of a few thousand years, implying that we must use fossil fuels more slowly than we do now if we are to avoid the "greenhouse limit" of 20 percent of supplies given above.

As an aside, it is likely that man's *past* activities have had a significant effect on the earth's carbon distribution. Specifically, in clearing land for agriculture and in obtaining fuel wood, man has reduced the forested portion of the earth's land surface from 60 percent to 30 percent. At present, the earth's land surface is roughly one-third forest, one-third grasslands or agricultural lands, and one-third barren (glaciers, deserts, etc.). The biomass per unit area of forests is twice that of grasslands and agricultural lands.[20] The indicated reduction in forested lands has lowered the terrestrial biomass roughly in a 4-to-3 ratio, that is, a reduction of about 400 billion metric tons of carbon. This compares with man's total fossil-fuel releases to 1970 of 140 billion metric tons. Because the land clearing took place over several hundred to a few thousand years, this release was fairly slow, and the deep ocean has had time to absorb a good portion of the carbon.

PART E / OTHER ENERGY-RELATED ENVIRONMENTAL PROBLEMS

In this part we lump together several environmental problems that do not fit into the three categories considered in Parts B through D. Damages to the environment occurring in the extraction, transportation, processing, and utilization of coal and oil are treated in the first two sections; damages associated with other energy sources are mentioned in the third section. Problems in the generation and transmission of electricity are treated in the fourth section. In the last section, we consider in a general way the ill-effects resulting from the end uses of energy.

6.21   Coal

Coal is extracted by surface mining (*strip mining*) and by underground mining (*deep pit mining*). There are substantial environmental damages resulting from both.

Land-Surface Damage

The standard procedure in the United States for *strip mining* level terrain is initially to remove the overburden from a narrow rectangular region and place the overburden along one of the sides (e.g., the left side). The uncovered coal is removed, and the operation next moves to the narrow rectangular region just to the right of the one just worked. Overburden from the new region is dumped into the hole created by working the old region. This process is repeated, moving to the right. The end topography is a series of ridges and valleys; the topsoil usually ends up on the bottom. If the terrain is hilly, *contour strip mining* is practiced. The excavation is made along a contour and the overburden is pushed downhill. Successive passes are made, working uphill. These two methods are illustrated schematically in Fig. 6.11. In both cases topsoil is lost, and the sub-surface overburden now on top is unlikely to support plant growth. Erosion is

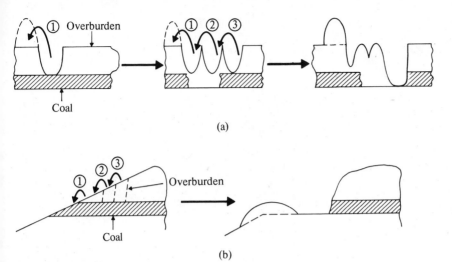

Fig. 6.11 Schematic illustrations of (a) area strip mining on level terrain, and (b) contour strip mining on slopes. Successive steps in moving the overburden are labeled ①, ②, ③.

severe, and on steep terrain landslides result. Nearby streams are damaged by siltation. Some examples of strip mining are shown in Figs. 6.12, 6.13, and 6.14.

The damages from strip mining can be reduced to some extent by reclamation procedures, i.e., grading, fertilizing, and seeding. However, the overburden which was in close proximity with the coal is very acidic and will not support vegetation. Unless it is well covered, reclamation may be only temporary. Excellent results have been obtained on flat terrain in western Europe, where the procedure is to handle the topsoil separately, returning it to the top. It is not clear that any procedure will work on steeply sloping terrain. On suitable terrain, cost estimates for proper reclamation range from 1000 to 10,000 dollars per acre and represent 5 to 10 percent of the price of coal.

Acid Mine Drainage

Pyrite (FeS_2) is an impurity invariably found with coal. It constitutes part but not all of the sulfur content of coal. If water comes in contact with pyrite, a dilute acid forms. Water passing through mines (underground or surface, active or abandoned) becomes acidic and subsequently damages streams. It is estimated [21] that 4 million tons of acid per year are discharged into streams in Appalachia (estimate for 1966), resulting in serious deterioration of 6000 miles of streams, roughly 4 percent of the streams in the area. Over 70 percent of the acid comes from underground mines, mainly abandoned mines.

Acid mine drainage from surface mines can be largely eliminated by good land reclamation procedure. Isolation of pyritic material is necessary both for revegetation and for elimination of acid drainage. Procedures for underground

Fig. 6.12 Area strip mining in Ohio. The large shovel removes the overburden; the small shovel removes the coal. Note size of trucks for comparison. (Photo courtesy of Consolidation Coal Company.)

mines have been less successful. Attempted methods include preventing water from entering the mine, removing it rapidly if it does enter, flooding abandoned mines if they are below drainage level, and sealing the entrances to abandoned mines. Treatment of the acid water itself is effective but expensive. It is appropriate if the water is to be used as water supply but rather steep for ecosystem protection, being in the neighborhood of one dollar per 1000 gallons.

Subsidence

When coal is removed by underground mining, the overburden tends to collapse into the cavity created by removal of the coal. By leaving occasional pillars of coal behind and by other techniques, the likelihood of a cave-in in an active mine can be kept fairly small. However, cave-ins in abandoned mines are common. The result is a settling or *subsidence* of all the overburden, with noticeable effects at the land surface above the mine. When subsidence occurs in an urban area (common in Pennsylvania), streets cave in and buildings are destroyed. In rural areas, the damage comes from upsetting surface and subsurface water-flow patterns. Frequently, subsidence enables water to flow into a mine, thus increasing acid mine drainage.

Fig. 6.13 Land appearance after area strip mining near TVA's Paradise Steam Plant, Kentucky. Note the characteristic ridge-and-valley pattern. (Photo courtesy of Billy Davis. Copyright © 1966, The Courier-Journal and The Louisville Times. Reprinted with permission.)

In the United States, about 6 million acres of land have been undermined in coal-mining operations, and about one-third of that has subsided. The only dependable, long-term measure to prevent subsidence is to backfill the mine, an expensive procedure.

Refuse Banks

Waste material is brought to the surface during the underground mining of coal. Cleaning of coal (in strip mining as well as underground mining) results in further solid wastes. (It also results in water pollution because the coal is usually washed.) Finally, when the coal is burned, the ash (typically 15 percent of the coal by weight) remains as a solid waste to be discarded. Invariably in all these cases, the material is merely piled in a heap. Past coal production has led to refuse amounts on the order of tens of billions of tons (enough to cover hundreds of thousands of acres to a depth of 100 feet). Current production (assuming 15

Fig. 6.14 A contour strip mine. Note the landslides. (Photo courtesy of the Tennessee Valley Authority.)

percent waste from extraction and cleaning, and 15 percent ash content) would yield one-sixth billion tons of waste per year in providing 25 percent of current U.S. energy needs. This would cover 1000 acres to a depth of 100 feet annually.

Refuse banks are unsightly and prevent land from being used for other more beneficial purposes. They are subject to erosion by wind, contributing airborne dust, and erosion by water, which spreads the waste material into surrounding areas. Rainwater leaches minerals from the refuse banks and damages nearby streams. The nature and magnitude of this damage depends upon the type of refuse. While ashes are alkaline, coal wastes are generally pyritic and contribute acid to streams.

The ideal solution to the refuse-bank problem would be to use the material to backfill old mines, solving the subsidence problem simultaneously. This is generally too expensive. Some refuse material can be used for construction purposes (cinder blocks, bricks, and roadbed material). With improved coal-burning techniques, some coal refuse heap material can be used as fuel.

Fires

There are a substantial number of fires in abandoned coal mines and in coal refuse banks. These are extremely difficult to extinguish, and frequently they are allowed to burn. A 1969 survey by the Bureau of Mines[22] reported 110 fires in mines and coalbeds and 460 fires in coal refuse banks. These fires contribute to air pollution and in some cases destroy a valuable coal resource. By damaging remaining coal pillars in abandoned mines, these fires increase the likelihood of subsidence.

Health and Safety for Coal Miners

Underground coal mining is a dangerous and unhealthy occupation. Accidents include cave-ins, fires, and explosions. Health effects are of a respiratory nature, principally black lung disease, which affects about 25 percent of coal miners. Stringent health and safety measures at coal mines can reduce these dangers— the Federal Coal Mine Health and Safety Act of 1969 is a step in this direction. Strip mining does not involve the above-mentioned occupational hazards, and also involves fewer workers per ton of coal removed. (The APHA study in the Related Reading list considers these hazards quantitatively on a damage-per-unit-energy-delivered basis.)

Summary

Can we give some semiquantitative characterization of the overall impact of coal extraction? The damages affect primarily Appalachia, a region of 100 million acres, or 5 percent of the U.S. land area. Within this region, damage is substantial. For example, 15 percent of the streams in West Virginia are contaminated by acid drainage. Erosion from strip-mined areas damages these and other streams. Strip mining directly affects one million acres in Appalachia (one percent of the land surface). Garvey[23] estimates the dollar damages attributable to coal mining to be about 3 billion dollars, and he stresses that intangible damages may be far higher. He estimates that a 10-percent increase in the price of coal would cover the tangible damages caused by new mining, and over a ten-year period would also cover those of past mining.

   The fraction of coal that is surface mined has been increasing over the past two decades and now represents about half of all coal being mined. This need not be a bad trend for the environment, since the damages from strip mining (on some types of level terrain) are believed to be avoidable at a price, and the price is probably less than 10 percent of the selling price of coal. Of course, the damages from strip mining are *not* currently being avoided, and very strong legislation will be required to correct this situation.

6.22   Oil

Use of petroleum for energy production can damage the environment during the extractive stage (leaks and blowouts at offshore wells), during transportation (tanker discharges, tanker accidents, pipeline problems), during refining (air and water pollution of various sorts), and during and after use (discharges from oil-fueled ships, evaporated fuel, discarded lubricants). Localized damage is caused to coastal aquatic ecosystems by high concentrations of petroleum. The possibility of damage to the ocean as a whole by long-term low concentrations is causing concern. Damage on land is less of a problem, in part because preventive and control measures are easier to apply. However, the trans-Alaska pipeline system (TAPS) is an exception and requires special consideration.

Coastal Oil Spills

Offshore wells now produce 17 percent of U.S. domestic petroleum, and this percentage is increasing. Some oil is released during normal well operations, but the real concern is over blowouts, such as occurred in the Santa Barbara Channel in 1969 and in the Gulf of Mexico in 1970. Substantial amounts of crude oil and petroleum products are moved by ship or barge, and accidents and spills connected with shipping are the other major sources of oil in high concentrations in coastal waters.

High oil concentrations along coastal regions foul beaches, kill marine birds, and can destroy breeding grounds of fish, shellfish, and other aquatic animals. It appears that the ecosystem damage caused by some of the spills has not been permanent, but ecological evidence indicates that near-shore areas differ widely in resistance to oil spills. The northeast coastline may be particularly vulnerable.

Oil in the Ocean

Estimates of direct release of petroleum into the world's oceans are given in Table 6.10. For comparison, world production of crude oil in 1969 was 1.8 billion tons. Unlike the coastal high-concentration problems, 90 percent of the oil entering the ocean is a result of normal operations rather than accidents.

Table 6.10 Estimates of direct losses of petroleum into the world's waters, 1969 (metric tons per year).

Source	Loss	Percentage total loss
Tankers (normal operations)		
Controlled	30,000	1.4
Uncontrolled	500,000	24.0
Other ships (bilges, etc.)	500,000	24.0
Offshore production (normal operations)	100,000	4.8
Accidental spills		
Ships	100,000	4.8
Other	100,000	4.8
Refineries	300,000	14.4
Rivers carrying industrial automobile wastes	450,000	21.6
Total	2,080,000	100.0

Source: *Man's Impact on the Global Environment,* Report of the Study of Critical Environmental Problems (SCEP). Cambridge, Mass.: MIT Press, 1970, p. 267. By permission.

The normal procedure for a tanker is to carry sea water as ballast when not carrying oil. This water is pumped into the ocean in deballasting and carries a residue of oil with it. This residue is typically 0.2 percent of the tanker's load of oil if no precautions are taken, but can be reduced a factor of 20 or more by *load-on-top* (LOT) deballasting. In this procedure, the oil residue floating on top of the ballast water is skimmed off and disposed of in port. Over 80 percent of the world's tankers practice LOT; the remainder are responsible for 24 percent of the oil discharged into the oceans. Further, since only one ship in eight is a tanker, and since almost all ships are fueled by oil and have oil tank cleaning, ballasting operations, etc., these other ships contribute substantially to oil in the oceans.

As discussed in Part D, petroleum products evaporate during processing and use, causing air pollution. The heavier hydrocarbon fractions form particles and settle out or are rained out onto the land or into the ocean. It is estimated that this source is perhaps four times larger than all the direct sources listed in Table 6.10. If so, the total amount of petroleum and petroleum products entering the ocean due to human activity is about 0.5 percent of annual production.

There is currently no firm evidence that any harm is being caused by low concentrations of oil in the ocean. However, the global nature of the situation requires extreme caution. It has been suggested [24] that petroleum in conjunction with chlorinated hydrocarbon pesticides (DDT and dieldrin) may be a dangerous combination. These pesticides are highly soluble in oil but are largely insoluble in water, and so will concentrate in oil. Since oil floats, this means that the pesticides will be spatially concentrated at the ocean surface. Measurements have shown concentrations at the surface 10,000 times larger than in the underlying water. Damage to species spending part of their life cycle near the ocean surface (e.g., plankton, larval stages of many fishes) and to their predators is a distinct possibility.

Alaskan Oil

In 1968, a large oil discovery was made on the Alaskan North Slope near Prudhoe Bay. Delivering this oil to markets presents a problem. The North Slope is ice-bound much of the year, so using tankers is precluded. (However, keeping a shipping channel open with ice-breakers has been considered.) Two pipeline routes were considered, one leading across Canada to the U.S. Midwest, and the other across Alaska to the ice-free port of Valdez for subsequent tanker shipment. The route to Valdez was chosen for reasons never satisfactorily explained.

Pipelines across land in the continental United States rarely present problems, but Alaska is a different matter. (1) The Brooks and Alaska mountain ranges must be crossed (see Fig. 6.15). (2) Much of the route is over permafrost (soil and rock whose temperature remains below 0°C); the friction from the flowing oil will heat the pipe to 140°F, and should this warm the permafrost, melting of ice would ensue and erosion problems and sagging of the pipe's foundation are likely. (3) The region is seismically active. (4) Many rivers must be crossed. For all these types of terrain, a pipeline break is far more probable than

Fig. 6.15 Country to be crossed by the Trans-Alaska Pipeline. This view into Dietrich Pass shows the wilderness characteristics, scenic beauty, and rugged nature of the terrain. (Photo courtesy of Wilbur Mills; reprinted with permission.)

under more conventional circumstances. (5) Photosynthetic activity in this region is low most of the year, so that damage to vegetation occurring either during the construction stage or from a pipeline break would be very slow to heal. (6) Erosion along the pipeline and service roads is expected. (7) The pipeline may constitute an obstacle to migration patterns of reindeer and other migratory animals.

The prime value of the land to be traversed by the pipeline is as wilderness, something of little value when there is lots of it, but of increasing value as it becomes scarce. It is a value very difficult to quantify, being of the psychological/ aesthetic or nature only varieties. The defining characteristics of a wilderness area are that it be large and that it be free of man-made features. The amount of wilderness remaining in the continental United States is small, and of low quality because most areas are near roads, are small in area, and are heavily used by people. Alaska, on the other hand, contains a large amount of high-quality wilderness—a large fraction of the state fits into this category. Since Alaska is the last remaining large wilderness area in the United States, it can be argued that preserving a major fraction of that wilderness should be a policy guiding any actions to be taken there. Such a policy directly conflicts with exploiting the oil resources of the North Slope.

This situation neatly demonstrates the inadequacy of conventional cost-benefit analysis. The benefit is evaluated to be 10 billion or more barrels of oil valued at, say, $4/bbl, yielding a benefit of exploiting North Slope oil of at least $40 billion. The cost is damage to land, the area of which is 800 miles (the length of the pipeline) by the width affected. This width might be considered to be as low as a few hundred feet, the directly damaged region, or perhaps as high as several miles, the region "less wild" due to the presence of the pipeline. Assuming an extreme width of 20 miles (10 miles on each side of the pipeline), the affected area is 16,000 sq mi or 10 million acres, about 3 percent of the land surface of Alaska. Wilderness land sells for less than $100 per acre, so a high estimate of the costs would appear to be $1 billion, which is much less than the benefits. By the same reasoning, the entire land surface of Alaska is worth less than $30 billion, and such cost-benefit analysis would mean that any benefit in excess of $30 billion would justify destruction of the entire land surface of Alaska. For perspective, the U.S. annual GNP is about $1000 billion. To destroy Alaska for a 3 percent increase in GNP for one year is utter nonsense! Clearly this is an inappropriate way to quantify the value of wilderness. The reason in part goes back to externalities, in this case external benefits. The benefits of wilderness accrue not just to the owner of the land, but to a much greater number of people. As a result, the selling price of wilderness land greatly understates its true value.

Coastal Land Use

In the United States, coastal land is at a premium. With 6000 miles of coastline and a total land area of 3 million square miles, the continental United States has only 0.2 percent of its land surface within a mile of the shore. This land is highly valued for industrial sites and for recreational development. Only recently has there been general recognition of the ecological importance of coastal marshes, estuaries, tidal wetlands, and the like, and of the role they play in ocean food chains. By now, much coastal land has been seriously degraded by human activity.

Offshore oil extraction entails the use of coastal land for collecting pipelines, storage tanks, and other onshore facilities. Utilization of imported oil entails the use of coastal land for docking facilities, oil storage tanks, and the like. For both offshore oil and imported oil, refineries are invariably located on coastal land. Increased pressure on coastal lands will result as an increasing fraction of U.S. oil supply comes from offshore and from overseas.

Summary

Past environmental damages associated with oil are far smaller than those associated with coal. Since oil resources will be exhausted long before coal resources (see Chapter 4), one could conclude that the cumulative damages from oil exploitation will be even smaller relative to cumulative coal exploitation damages.

This conclusion is probably correct. However, we must be watchful for an increase in damages as a resource becomes depleted. This trend is already very evident where oil is concerned in the United States. As oil reserves in safe environments disappear, the bulk of the exploration and production shifts to more hazardous environments (i.e., offshore areas and Alaska) where accidents are both more likely and more damaging. Further, depleted resources implies more transport of oil (e.g., more imports), so transportation dangers increase. As domestic oil production declines significantly (as someday it must), desperate actions with severe environmental consequences may well be forthcoming. It is possible that the decision to exploit oil shale reserves is such an action (see Section 6.23).

6.23   Other Energy Sources

Natural Gas

Compared to coal and oil, natural gas is relatively free of environmental problems. It is transported under high pressure in pipelines, and so the possibility of a pipeline break in a populated area represents a slight safety hazard. Similarly, LNG is stored in populated areas, and a failure of a storage tank (possible but unlikely) would endanger many lives. Gas leaks in homes have led to explosions and fires, though this is rather infrequent now. Natural gas extracted in Alaska will entail another pipeline across Alaska and/or Canada, and many of the remarks about Alaskan oil are applicable (see Section 6.22). Similarly, offshore natural gas extraction entails the use of coastal land, and the remarks of Section 6.22 on coastal land use apply here also. The most significant environmental problem associated with natural gas is resource depletion—there isn't enough of it.

Shale Oil

The Department of the Interior recently decided to lease oil shale land (the most favorable deposits are on public land), and has completed the environmental impact statement[25] as required by NEPA. Four leases were sold during 1974. The intention is that this be a prototype program to be evaluated before additional leases are sold. The EIS was written in terms of the impact made by a shale-oil industry producing one million bbl/day. (The four sites leased so far might produce about one-fourth of that figure.) In comparing environmental costs against economic benefits, we should note that one million bbl/day is 6 percent of U.S. oil consumption in 1972, or 2.4 percent of U.S. energy consumption for that year. That is not a negligible amount, but it is pretty small.

The impact area in question, near the junction of Utah, Wyoming, and Colorado, is presently characterized by low population density, negligible industrial development, high air quality, and short water supplies. All of these factors speak against shale-oil development.

The dominant feature of a shale-oil operation is materials handling—mining and waste disposal. Even with the richest deposits, 85 percent of the material mined is waste. This means that per unit of energy extracted, shale oil requires about 5 times more material to be mined than does coal, and about 17 times more waste material to be discarded. Since coal mining is not an environmental delight (see Section 6.21), we have reason to expect that a large-scale shale-oil industry would be an environmental nightmare.

Both surface mining and underground mining of oil shale are comtemplated, and problems of land surface degradation, subsidence, and miner safety are to be expected. With the large amount of spent shale to be discarded, backfilling mines (both surface and deep pit) will receive more consideration. This can also help solve the subsidence problem. However, the volume occupied by the spent shale is larger than the original shale by at least 12 percent due to empty spaces in the processed shale. As a result, some surface disposal will be necessary. (One proposal is to use spent shale to fill "useless" canyons. This led one environmentalist to note that he was unaware that there were any useless canyons.) The EIS projects that a 30-year operation of a one-million-bbl/day industry would affect a land surface area of 50,000 acres without backfilling, and 35,000 acres with backfilling.

Some components of spent shale are water soluble and make the water highly alkaline. Specifically, there are high concentrations of calcium, sodium, and potassium sulfates. Surface runoff must be controlled to prevent erosion and to stop leaching of surface minerals until vegetation stabilizes the surface. But plants will not grow in shale unless it is leached, fertilized, mulched, and irrigated. Water is a necessary ingredient for land-surface reclamation generally, and in a region where water is scarce, there will be temptation not to "waste" it in this way.

A shale-oil industry will adversely affect the water of the area in a variety of ways. (1) The mining operations will change ground-water patterns, both because of the physical changes made underground and because of pumping done to keep ground water out of mines. (2) Both surface water and ground water will leach minerals from spent shale. (3) The industry itself will require water in substantial quantities, about three gallons of water for each gallon of oil produced. A one-million-bbl/day industry would consume half the estimated potential water supply of the immediate area. Technological improvements can reduce the water requirements per unit of oil produced, but it is unlikely that production could exceed a few million bbl/day with locally supplied water. (4) The increase in population and commercial and industrial activity caused by a shale-oil industry will make additional demands on water supply.

Let's look at the water situation more quantitatively. The discharge of the Colorado River, with a 0.25 million sq mi drainage basin (which includes the oil-shale region), averages 6000 cfs. The total discharge from all rivers of the U.S. Southwest (1 million sq mi) is 60,000 cfs. Discharge for the entire continental United States (3 million sq mi) is 1.8 million cfs. The discharge of the Colorado River, per unit of drainage basin area, is a factor of 25 below the U.S. average, and the discharge from the U.S. Southwest is a factor of 10 below average. This quantifies the statement that water is in short supply in the oil-shale area. A one-million-bbl/day shale-oil industry requires 3 million bbl/day or 200 cfs of water, which is 3 percent of the Colorado River discharge. Removal of this water will result in increased salinity of the lower Colorado of 2 percent, with an economic loss to the lower Colorado River basin estimated in the EIS of a million dollars a year (small compared to the value of the oil obtained). Minerals leached from spent shale would increase the salinity further. The EIS claims that this can be controlled and would result in negligible increase, but I have my doubts. Summarizing, a small, one-million-bbl/day shale-oil industry will cause noticeable changes in the water situation of the entire river basin (8 percent of the U.S. land area) and cause major changes in the water situation in the immediate area of the industry. (It is worth noting parenthetically that any energy development has some water requirement. In particular, the development of Rocky Mountain coal resources will be limited by water supply.)

A shale-oil industry will add dust and particulates to the air as a result of mining, crushing, retorting, and spent-shale handling operations; it will add sulfur dioxide to the air as a result of the retorting operation. The expectation is that air quality will decline perceptibly but still be above state and federal air-quality standards. In a region of very high air quality, one can argue as to whether it is wiser to make use of the atmosphere's waste-disposal capability (i.e., dirty up the air) or to retain the advantages of "unnecessarily" clean air.

Finally, we must consider the change in the regional lifestyle that a shale-oil industry implies. It is estimated that a one-million-bbl/day shale-oil industry would result in a doubling of the area population, from 100,000 to 200,000. (This estimate includes commercial activities (stores, schools, etc.) necessitated by the industry but does not include other industrial activity which might be attracted by the shale-oil industry.) At present, the region retains its natural character because there is a lack of people and industry. A sizable shale-oil industry would change this and move the region toward urbanization.

In assessing the overall environmental impact of a shale-oil industry, it is useful to draw comparisons with coal. The areas adversely affected by coal extraction considerably exceed the land area directly damaged. (In Appalachia, only about one percent of the land has been strip mined and about 4 percent of the land has been deep mined, yet a large fraction of Appalachia has been degraded.) Damage is extended beyond the immediate mine area by erosion and by acid mine water. In the case of shale-oil exploitation, the extending factors will be erosion, spent-shale leaching, and water consumption. Because of the sizable increase in materials to be handled, it appears likely that the environmental damage per unit of energy obtained will be greater in the case of shale oil than in the case of coal, assuming equal effort is made to minimize such damage. A large-scale oil-shale industry capable of supplying a significant fraction of U.S. energy needs appears to me to be a bad idea.

Tar Sands

Tar sands are similar to oil shales in that large volumes of material must be surface mined and then discarded in order to obtain relatively small amounts of energy.

Geothermal Energy

Water coming to the surface in geothermal steam wells usually has a very high mineral content and would pollute streams if released to them. Often it can be reinjected into the geothermal area to replace removed water.

Solar Energy

The principal environmental effect of a large terrestrial solar-energy installation would be the large amount of land it would occupy. By increasing the absorptivity of the area, and by sending the energy as electricity to other regions, the facility would slightly alter the earth's energy balance and energy distribution.

Nuclear Energy

The environmental ill effects of fission energy are covered in Section 5.2. Until a working fusion-energy scheme is available, we can only surmise as to what its environmental ill effects will be. Assuredly there will be thermal pollution; the amount will depend on the efficiency of conversion from nuclear to electrical energy. The quantity of radioactivity produced will be considerably less than for fission, but it does not follow that releases to the environment will necessarily be smaller. (A ^{235}U fission releases 200 MeV, creates two fission-fragment decay chains, and liberates 2 to 3 neutrons. Fusion reactions releasing 200 MeV would liberate 10 to 16 neutrons (see Eqs. 5.23 and 5.24). Thus while fusion does not create radioactivity directly, it liberates about 5 times more neutrons per unit of energy released. Some fraction of the neutrons, upon capture, will create radioactive isotopes. The amount of radioactivity created and the amount released to the environment will depend on the design details of the fusion device. Recall that neutron capture on impurities in the cooling water of a fission reactor is a significant component of radioactive releases, and care is taken to eliminate these impurities.) Other environmental problems will depend on the specifics of the scheme, but are believed (hoped?) to be small.

Environmental problems associated with hydroelectricity as an energy source are treated in Section 6.24.

6.24   Electricity

Steam-Electric Power Generation

Most of the environmental problems associated with steam-cycle generation of electricity have already been discussed in those sections dealing with thermal pollution, radioactivity, and air pollution (Chapter 6, Parts B through D). Ash disposal was treated in Section 6.21. The use of cooling water involves problems beyond those of thermal pollution, and we mention some briefly here.

The large volume of cooling water passing through the condenser may carry with it a wide variety of organisms (algae, zooplankton, small fish, etc.), and they may be killed by the thermal shock on passing through the condenser, by changes in pressure, or by mechanical damage. Intake screening intended to prevent larger organisms from entering the cooling water flow may itself damage fish or invertebrates. Biocides (e.g., chlorine) used to control fouling of condenser tubing may kill entrained organisms or organisms near the cooling water discharge area. If the cooling water intake and discharge are at different levels, the characteristics of one level will be transferred to the other with possible ill effects. For example, if the intake is located in deep water and the discharge is near the surface, the discharged water will be higher in nutrients and lower in dissolved oxygen than the surface water.

Hydroelectricity

The majority of hydroelectric facilities involve constructing a dam across a river to create a storage reservoir and an abrupt change in water elevation. This results in a major change in the area: a river valley is transformed into a lake. One can debate as to whether this is an improvement or a deterioration of the environment. After all, there are recreational and psychological/aesthetic values to lakes, just as there are to free-flowing rivers. But in the United States, there are many lakes and reservoirs, whereas there are few remaining rivers that are both free from dams and free from extensive human developments (i.e., cities). The conditions requisite for a good dam site seem to ensure choice of a location with high aesthetic value. The area to be flooded must contain few buildings (too expensive if developed land must be purchased), and therefore is invariably in a wilderness or semiwilderness area. The need for vertical drop results in choice of a wild, tumultuous river rather than a placid one. The need to minimize the length of the dam results in the choice of a narrow valley, canyon, or gorge, rather than duller, flatter terrain.

It is illuminating to list some dams and proposed dam sites to illustrate the above. The Hetch Hetchy Valley, located in Yosemite National Park and described as being of beauty comparable to Yosemite Valley, was authorized as a dam site by Congress in 1913 over strong opposition from conservationists, among them John Muir. This was probably the first major "dam battle," and many more have followed. The Federal Bureau of Reclamation has plans for dams along the Colorado River that would turn it into a series of reservoirs, including that portion of the river in the Grand Canyon. Much of the Sierra Club's rise to national prominence was due to its leadership in successful opposition to dams in the Grand Canyon. The Glen Canyon dam, upstream from the Grand Canyon, was little opposed by conservationists, in part because the area was so remote that few people were aware of its beauty. In the populous Northeast, wilderness is practically nonexistent. The only two large regions relatively devoid of roads are northern Maine and the Adirondack region of New York. Both areas have had dam proposals in recent years. The proposed Gooley Dam across the upper reaches of the Hudson River would have inundated 16,000 acres of Adirondack forest and involved a stretch of the river famous for white-water canoeing. In northern Maine, there have been proposals to dam the Allagash and the St. John, two rivers legendary from logging and trapping days. For both rivers, the dam would destroy essentially the entire river above the dam as well as a large area of forest. At the moment, the Gooley Dam project is quite dead, the Allagash River has been protected as a park, but the project for the St. John has been revived. Known as the Dickey-Lincoln Dams, the project would flood about 90,000 acres of woodland. It could produce a peak power of 760 MW_e, but a daily average power of only 125 MW_e, and would be used for peaking rather than base-line supply. Many people think 90,000 acres of wilderness and a free-flowing river is too steep a price to pay for 125 MW_e. A major dam battle is in the making.

Dams represent an obstacle to migrating fish (such as salmon) that is only partially alleviated by the use of fish ladders. Variations in flow rates caused by the installation are damaging to the riverine ecosystem downstream, and at the same time variations in lake levels are damaging to the lake ecosystem. Entrainment of organisms in the water driving the turbine results in damages similar to those described for cooling water at the beginning of this section. One can continue in this vein to list many ways in which the new ecosystem created by the dam may not function in an optimum way, and also list a variety of remedial steps.[26]

We have raised two somewhat distinct types of objections to hydroelectric facilities. (1) The new ecosystem may not function as perfectly as we would wish. (2) Even if the new ecosystem does function smoothly, the former ecosystem may have been preferable. One can discuss the first type of objection in a scientific and objective way (i.e., with a dam, with fish ladders, by what fraction the salmon population will be reduced over its free-flowing river value, etc.). The second type of objection lends itself to more emotional reactions, as the first two paragraphs of this subsection surely indicate. Yet I believe in almost all cases this objection is far more important than those of the other sort. The real objection to Hetch Hetchy Reservoir is not that its level fluctuates, but that it has destroyed Hetch Hetchy Valley. The Grand Canyon of the Colorado River is far more desirable than the Grand Canyon of Lake Who-Knows-Who. If these things are so obvious, why can't we develop the scientific discipline to enable us to discuss them in an objective, quantitative way?

Most dam projects are proposed not solely for hydroelectric power, but also for flood control, recreation, and water supply. It is important to realize that these are *competing uses* rather than compatible uses. Flood control is provided by that part of the reservoir that is *not* filled (i.e., the air in the reservoir) and argues for keeping the reservoir as near empty as possible. Water supply and hydroelectricity utilize the water in the reservoir and argue for keeping the reservoir as near full as possible. Recreation requires a constant water level (to avoid damage to fish and to avoid exposure of unattractive submerged shoreline).

Pumped Storage

Electricity is not used at a steady rate. There are variations during the course of a day, from weekday to weekend, and from summer to winter. Since installed generating capacity is constant, this means that some capacity will be idle some of the time. Pumped storage is a scheme to help level out the daily load. During the night hours when demand is low, electricity from baseline generating plants is used to pump water from some low point (e.g., a river) to some high point (e.g., an artificial lake on top of a hill, typically 400 to 1000 feet higher). During peak hours, this water is allowed to flow back downhill, driving turbines which drive generators in the process. In this way, electricity generated during off hours is stored for use during peak hours. The efficiency of the process is about 70 percent, that is, for every 3 kWh of electricity used to pump water to the upper reservoir, about 2 kWh of electricity is recovered as it flows back down. The economics are such that the inefficiency in the storage process is more than compensated for by the lower costs of the pumped-storage facility as compared to additional conventional capacity.

Because the storage efficiency is less than 100 percent, the environmental problems associated with the original generating plant are increased. The method produces 50 percent more air pollution, thermal pollution, etc., in generating 2 kWh via pumped storage as it would generating it directly. For pumped storage, a site is required with a large elevation change and a ready supply of water, near a load center. This translates into a hill by a river near a city. In some cases these sites have considerable aesthetic and recreational value which the pumped-storage facility may damage. The pumphouses, pipes, and transmission lines may constitute a visual intrusion. The water level of the upper reservoir undergoes large daily fluctuations—50 feet is common—so it has no recreational value. Aesthetically it is an eyesore. There are concerns about entrained organisms in the water as there are for hydropower, but now the likelihood of damage is multiplied by 2 because these organisms must survive the pumps on the way up and the turbines on the way down. If the flow to and from the upper reservoir is not small compared to the natural flow of the lower reservoir (here assumed to be a river), any species depending on river flow for cues for migratory movement may be adversely affected.

Certainly the most famous pumped-storage facility is one that has not yet been built. It is Consolidated Edison Company of New York's project on Storm King Mountain above the Hudson River. Proposed by Con Ed in 1963 and approved by the Federal Power Commission in 1965, the matter was taken to court by Scenic Hudson Preservation Conference (a coalition formed to oppose the project). Scenic Hudson charged that the FPC had not considered the environmental effects (specifically, the scenic values at the mountain and around the reservoir, and damage to fish) in granting a building permit. The U.S. Court of Appeals ruled in favor of Scenic Hudson and required FPC to reconsider, taking environmental factors into account. Con Ed modified its plans somewhat, FPC reconsidered, and in 1970 again issued a building permit. Again the matter went to the courts, which in 1971 ruled that FPC's approval was proper. Undaunted, Scenic Hudson continued the battle in the courts and as of now, the matter is still not settled. In December 1973, a federal court ruled that Con Ed must obtain a permit from the Army Corps of Engineers for dredging operations. Now under NEPA, this will require a full environmental impact statement.

Transmission of Electricity

There are presently about 300,000 miles of power lines in the United States, with rights-of-way occupying about 4 million acres. By 1990 it is projected that 200,000 more miles of lines will be added, bringing the total acreage to 7 million (one-third percent of the U.S. land area). The damage these transmission lines present is largely aesthetic—the lines and towers don't look pretty, and in some situations they mar otherwise very attractive landscape. And the trend is toward higher transmission voltages, which means larger tower structures and more visual intrusion.

The visual impact of transmission lines and towers can be reduced by care in routing, for example, avoiding the crest of a hill where the towers silhouette against the sky. The basic tower unit can be designed with more consideration given to aesthetics, so that they have some semblance of grace.

An increasing number of power distribution lines are being placed underground. This can also be done for transmission lines, but installation is very expensive. Currently it is done only in dense urban areas (where the right-of-way for above-ground transmission would be even more expensive) and where there are compelling aesthetic reasons. Technological advances may reduce the cost of underground transmission, and efforts in this direction are being encouraged.

6.25   Deliberate Consequences of Energy Use

So far we have been considering the indirect, undesired consequences of energy use. Now we must give some thought to the direct, desired consequences. A few examples will clarify the distinction. In burning oil to heat a house, the indirect consequences are the problems associated with the extraction, transportation, and refining of oil, and the contributions to air and thermal pollution when the oil is burned. The direct consequence is that the house is warmed. In driving a car from home to work, an indirect consequence is the air pollution produced; the direct consequence is that the individual moves from home to work.

The direct, desired consequences of energy use are not always environmentally benign. For example, filling wetlands and building roads are desired consequences that impact unfavorably on nature. It is my contention that *even if all undesired consequences of energy use could be eliminated, at some level of energy use per unit area, the* desired *consequences would result in excessive environmental damage.*

In a very rough sense, the environmental damage from the desired consequences of energy use, the fractional "upsetting of nature," is proportional to the energy used per unit area. Consider the desired result of home heating. If only half as much energy were to be used, houses would be smaller and duplexes and apartments more common, with the result that a smaller amount of land would be used for housing. Consider transportation to and from work. A smaller energy use could result by living closer to work (reducing suburban sprawl and the amount of road surface), car pooling (reducing the amount of road surface), relying on mass transit (less sprawl, fewer roads), using smaller cars with better gas mileage (less environmental impact in the mining of iron and the manufacture of the automobiles). Consider the desired end result of manufacturing various products. Less energy available would assuredly result in reduced production per year, with consequent reduced environmental impact. In every case that one considers, had less energy been available for a given desired end result, the area would most probably have developed in a fashion making less extensive or less severe intrusions on natural ecosystems. It must quickly be admitted, however, that human material well-being, *as conventionally measured,* also would have been lower.

Another way of stating the above is to note that the extent of intrusions into natural ecosystems in some area is roughly proportional to the amount of human activity in that area, and energy use is a fairly good measure of human activity. Less energy implies less human activity; less human activity implies less disruption of nature. Following this to its logical conclusion, no energy implies no human activity and no disruption of nature. The conclusion is certainly valid, but not therefore an end toward which to strive. (Even though it would likely be better for all species other than *Homo sapiens,* this book is written by and for members of that species, and so *Homo sapiens* receives special consideration!) The question might more properly be stated: How large a level of human activity, as measured by energy use, can result in a disruption of natural ecosystems that is still acceptably small?

Different energy uses have quite different environmental impacts per unit energy. Energy used for warming the atmosphere causes considerably less damage than the same amount of energy used for running chain saws, which in turn causes considerably less damage than the same amount of energy used for starting forest fires. In the ensuing discussion, we assume that energy is not being used to deliberately damage natural ecosystems, but is being used judiciously to accomplish desirable human ends.

As a rough rule of thumb, we would expect human influences or natural influences to predominate in a region in proportion to their relative energy uses. If natural energy uses greatly exceed human energy uses, man's presence will be benign, whereas if human energy uses exceed natural uses, man will dominate and ecosystem function will be altered. Since we are talking about "living nature," the appropriate measure of natural energy use is photosynthetic activity. For "physical nature"—weather and climate—the appropriate measure is total energy flow.

Globally, total energy flow is $173,000 \times 10^{12}$ watts, which averages to 350 watts/m^2. Global photosynthetic activity is estimated to lie between 30 and 200×10^{12} watts, with a best guess being 120×10^{12} watts. This averages to about 0.25 watt/m^2. Photosynthesis on land considerably exceeds ocean photosynthesis, terrestrial photosynthesis averaging 0.5 watt/m^2 and ocean photosynthesis averaging 0.1 watt/m^2. Human energy use is currently (1970) 6×10^{12} watts, and takes place essentially all on land, yielding a terrestrial average of 0.04 watt/m^2.

On a global terrestrial basis, human energy use is about one-tenth that of photosynthesis, so we would expect that man's presence is apparent but not dominating. This is in accord wtih our experience. In consulting Table 6.2, we can consider some smaller regions. For urban areas, human energy use greatly exceeds photosynthesis, and we would expect that natural ecosystems are greatly altered, even perhaps totally destroyed. For larger regions such as the eastern United States or western Europe, human energy use and photosynthesis are comparable, and we would predict that natural ecosystems are noticeably altered, that human activity is much in evidence, but that nature is not overwhelmed. All this agrees with our experience, suggesting that the comparison of human energy use with photosynthesis is a valid approach as long as one uses it only as a rough guide.

One might argue that the preceding merely shows the effect of human population density on natural ecosystems, since energy use per capita is roughly constant for industrialized nations. We might look at Asia for some comparisons. There the per-capita energy use is much lower, but population densities are high. No clear answer is forthcoming. There are bad agricultural practices (e.g., overgrazing) caused by the shortage of food, and absence of pollution abatement due to the low standard of living. These we might attribute to too little energy per capita, rather than too much. Generally, however, we must conclude that the environmental ill effects caused by the average Indian, Pakistani, etc., are much lower than those caused by the average individual in an industrialized country.

In any case, the real question is not whether it is possible to damage the environment with a high population density and a low energy use, but whether it is possible to avoid damage to the environment with a high energy use independent of the population density. While making no claim to having proven it, I strongly suspect that as human energy use per unit area begins to dominate photosynthetic activity, ecosystem alteration becomes unavoidable.

Let us imagine an increase in human energy use of a factor of 10, making human energy use and terrestrial photosynthesis about comparable. This would mean that the average land area would be more "manipulated" than the United States is now. If U.S. energy use increased a factor of 10, the average U.S. land would be like Bos-Wash, as would western Europe. This seems too much to me.

Questions and Problems

6.1
Consider several examples of environmental damages (e.g., from Table 6.1). For each, explain the nature of the damage and explain how it does or does not represent an external cost. In each case of an external cost, suggest a scheme for internalizing the externality and discuss the problems and advantages associated with your scheme.

6.2
a) Cooling water flows through a power plant at a rate of f cfs. Waste heat is added to the cooling water at a rate of P MW. Derive a formula for the temperature rise $\Delta T°C$. For $P = 2000$ MW and $f = 2000$ cfs, determine ΔT.

b) When waste heat is removed from a power plant with once-through cooling by means of a lake or cooling pond, about half of the energy eventually enters the atmosphere in the form of latent heat via water evaporating from the lake or pond. Derive a formula giving the rate of evaporative water loss e (in liters/sec or cubic feet/sec) in terms of the rate of heat removal from the power plant P (in MW). For $P = 2000$ MW, evaluate e.

6.3 *

a) A river of width w, depth d, and flow rate f has heat energy added at a rate P by a power plant. This causes the temperature of the river at a distance x downstream to be $\Delta T(x)$ above its natural temperature. By obtaining and solving a differential equation for $\Delta T(x)$, show that $\Delta T(x) = \Delta T(0)e^{-x/x_0}$. Give expressions for $\Delta T(0)$ and x_0. (In your derivation, assume (1) the temperature is uniform over the width and depth of the river, namely, there is complete vertical and lateral mixing, and (2) the temperature derivative of the surface energy transfer, dE_{ex}/dT_s (as in Eq. 6.5), may be treated as a constant.)

(b) Though there is considerable variation from river to river, the relationships between flow, width, depth, and velocity for "average" rivers and streams are as shown below:

Flow (ft³/sec)	Width (ft)	Depth (ft)	Velocity (ft/sec)
100	50	1.0	2.0
1,000	200	2.3	2.2
3,000	350	3.4	2.5
10,000	600	6.7	2.5
100,000	1,200	28.0	3.0

Source: Leopold, L. B. et al., *Fluvial Processes in Geomorphology*. San Francisco: W. H. Freeman, 1964.

Evaluate x_0 for all flows given. Evaluate $\Delta T(0)$ for a flow of 10,000 cfs and a power addition of 2000 MW. A reasonable value to use for dE_{ex}/dT_s is 30 watts/m²-°C. If vertical mixing is incomplete, and the warm return water floats on the cold river water, how will $\Delta T(0)$ and x_0 change?

6.4

If the thermal conductivity of water were high, the temperature profile of a thermally stratified lake (Fig. 6.2) would be "washed out" by heat crossing the thermocline by conduction. Show that the thermal conductivity of water is sufficiently small that this does not happen. Specifically, consider a thermally stratified lake with a hypolimnion at a temperature of 5°C and an epilimnion at a temperature of 20°C, separated by a thermocline 2 meters thick. Calculate the heat transferred across the thermocline in a 6-month period. (The thermal conductivity of water is 1.5×10^{-4} kcal/(sec-m²-°C/m).) If this heat were uniformly distributed over the hypolimnion, assumed to be 30 meters deep, how much would the hypolimnion temperature rise? If instead the heat were concentrated in the upper portion of the hypolimnion, raising its temperature to that of the epilimnion, what depth of hypolimnion would be converted to epilimnion?

(The reader with more background in mathematics and physics may prefer to work with the heat diffusion equation for initial conditions of an epilimnion 10 meters deep at 20°C in contact with a hypolimnion 30 meters deep at 5°C, and see how the temperature profile develops over a 6-month period.)

6.5

For a lake to become and remain thermally stratified, the buoyant force due to the density difference between warm and cold water must be stronger than the forces from wind and wave action that tend to mix the lake waters vertically. Near the surface of a lake, forces due to winds and waves are strong, and the top several meters will always be mixed. These forces become much weaker with increasing depth, soon dropping below that due to the density difference. In this problem, investigate the magnitude of the buoyant force.

a) Plot the density of water vs. its temperature from 0°C to 30°C (data are in *The Handbook of Chemistry and Physics*). Note the temperature of maximum density and those temperatures at which the density has decreased by 0.01 percent, 0.05 percent, 0.1 percent, and 0.2 percent.

b) Empirically, lakes become stratified with epilimnion temperatures of 12°C and hypolimnion temperatures of 4 to 6°C. What percentage density difference does that correspond to? With this density difference, how does the buoyant force on a volume of epilimnion submerged in hypolimnion compare with the gravitational force on the same volume? How does the pressure required to submerge a 10-meter depth of epilimnion into the hypolimnion compare with atmospheric pressure? (In both cases, you will find the buoyant force or pressure relatively small, but the particuar comparisons are not relevant to the problem of stratification in a lake.)

c) Assume a volume of epilimnion with density ρ, horizontal cross-sectional area A, and height h is drifting downward through the surrounding epilimnion due to momentum acquired from wind and wave action (or any reason whatever), and has a velocity v as it encounters the hypolimnion where the density is ρ_0. (For simplicity, the thickness of the thermocline is assumed to be zero.) How deeply will the volume of epilimnion penetrate the hypolimnion before the buoyant force due to the density difference stops and reverses the downward motion? Evaluate numerically assuming that $v = 5$ cm/sec (a fairly high value for vertical currents in a lake), $(\rho_0 - \rho)/\rho_0 = 0.001$ (a typical value), and $h = 5$ m.

6.6

Estimate the amount of oxygen reaching the hypolimnion by diffusion across the thermocline with the following model. A well-mixed epilimnion with oxygen concentration ρ_0 is separated from a well-mixed hypolimnion with oxygen concentration ρ and depth d by a thermocline s meters thick. Assume the oxygen concentration varies linearly across the thermocline from ρ_0 to ρ. Assume ρ_0 is constant in time, and $\rho = 0$ at $t = 0$. Calculate $\rho(t)$ for later times t, for $\rho(t) \ll \rho_0$, in terms of the diffusion constant D and the quantities given above.

Evaluate $\rho(t)/\rho_0$ numerically for $s = 2$ meters, $d = 30$ meters, and $t = 0.5$ year. (The diffusion constant for O_2 in water is 2.1×10^{-9} m²/sec.) Does your result justify the assertion that the amount of oxygen reaching the hypolimnion by diffusion across the thermocline is negligible, and the hypolimnion must therefore get its oxygen by vertical mixing when the lake is not stratified?

6.7

Consider three alternatives to once-through cooling for a 1000-MW_e nuclear power plant with 2000 MW of waste heat to handle.

a) For a cooling pond, how large a pond is required? What flow of makeup water is required to replace that lost through evaporation? (As for natural lakes, about half of the energy transferred to the atmosphere from a cooling pond takes place by evaporation.)

b) For an evaporative cooling tower, what flow of makeup water is required to replace that lost through evaporation? (Essentially all the energy transferred to the atmosphere from an evaporative cooling tower takes place by evaporation.) What flow of air is required to evaporate that volume of water? (Assume air enters with a relative humidity of 40 percent and leaves with a relative humidity of 90 percent.) In addition to evaporation, water is also lost by "drift" (some liquid droplets are carried away by the airflow) and by "blowdown" (minerals in the circulating water are concentrated by evaporation, so the water must be periodically discarded to get rid of them). Assuming drift is 15 percent of evaporation and blowdown is 35 percent of evaporation (reasonable values), what is the total flow of makeup water required?

c) For a dry cooling tower, if energy is transferred to the atmosphere by warming air $15°C$, what flow of air (at stp) is required? Assume this air flows through ten towers each having a base diameter of 100 meters. With what speed does the air move in the lower portion of the towers?

6.8 * ‡

Construct a model for a lake, using the energy transfer equations 6.1 through 6.3. For climatic conditions, assume the following. (1) The radiant energy absorbed by the lake E_s varies sinusoidally over the course of a year: $E_s = E_0 + \epsilon_s \sin 2\pi t$, t being measured in years. Take plausible values for E_0 and ϵ_s. (2) The temperature of the atmosphere T_a varies sinusoidally over the course of a year, with a phase lag behind the radiant energy: $T_a = T_0 + \tau_a \sin (2\pi t - \phi)$. Take plausible values for T_0, τ_a, and ϕ. (3) The partial pressure of water vapor in the atmosphere p_a is $\frac{1}{2} p(T_a)$, where $p(T_a)$ is the vapor pressure of water at a temperature T_a. That is, assume the relative humidity of the atmosphere is 50 percent. (4) The wind speed W is constant at 4 mph.

Empirical information you may find useful includes the following. (1) $\kappa = 0.26$ mm Hg/°F (used in Eq. 6.3). (From Asbury, Ref. 9.) (2) $f(W) = 11.4$ W Btu/ft²-day-mm Hg, W being measured in mph (used in Eqs. 6.2 and 6.3). (From Asbury, Ref. 9.) (3) $p(T) = 4.5 + 0.32T + 0.02T^2$ mm Hg, T being measured in °C (used for $p_s(T_s)$ and $p(T_a)$ in Eq. 6.2 and assumption (3) above). This formula is a rough fit by the author to data in *The Handbook of Chemistry and Physics,* valid for $-5°C \gtrsim T \gtrsim 30°C$. You can easily improve upon this fit and extend its range by taking more terms.

a) With the above assumptions and empirical information, you can obtain an expression for the energy transfer across the lake surface, E_{ex}, as a function of t, the time (in years), and T_s, the temperature of the lake surface. Be careful to use a consistent set of units!

b) Assume the lake consists of a well-mixed epilimnion of temperature T_s and depth d_e, separated from a well-mixed hypolimnion of temperature T_h and depth d_h by a thermocline of negligible thickness. Assume the lake is thermally stratified for $T_s > 6°C$ and $T_s < 3°C$. For $3°C \gtrsim T_s \gtrsim 6°C$ the lake is not stratified, complete vertical mixing occurs, and $T_s = T_h$. Thus the lake as a whole warms to 6°C, and then the epilimnion warms further while the hypolimnion remains at 6°C. Similarly, the epilimnion cools until it reaches 6°C, the lake as a whole cools from 6°C to 3°C, and then the epilimnion cools further while the hypolimnion remains at 3°C.

With these assumptions, derive the differential equation for T_s:

$dT_s/dt = -\alpha\, E_{ex}(T_s,t)$,

where α is a "constant" with two values, one for $3°C \gtrsim T_s \gtrsim 6°C$ and another for other values of T_s,

c) Insert the expression for $E_{ex}(T_s,t)$ found in (a) into the differential equation derived in (b) and solve in some way (being sure your units are consistent). Probably numerical integration with a computer is the best method of solution. Investigate the sensitivity of the solution to the initial conditions.

d) Given a dependable method for finding $T_s(t)$, investigate the sensitivity to the various parameters of the problem (d_h, W, T_0, etc.). From $T_s(t)$, determine the duration of thermal stratification. Does the lake freeze in the winter? Pick some specific lake, modify the climatic assumptions to correspond to those for the region, and compare your results with data for the lake. (This "problem" can easily be expanded into a term project.)

6.9

a) Plot vs. time the number of decays per second from a sample initially containing 1 curie of ^{131}I (8-day half-life), using 3-cycle semilog paper. On the same scale, plot the number of decays per second from an initial sample of 0.1 curie of ^{89}Sr (51-day half-life). Cover a 100-day period of time.

b) Compute and plot as a function of time the number of decays per second from a mixture initially containing 100 Ci ^{133}I (21-hour half-life), 10 Ci ^{131}I (8-day half-life), 1 Ci ^{89}Sr (51-day half-life), 0.1 Ci ^{134}Cs (2-year half-life), and 0.01 Ci ^{137}Cs (30-year half-life).

c) Calculate the total number of decays from each variety of radionuclide for the mixture given in (b) above. Derive formulas for the total number of decays from a sample initially containing S curies of radionuclide with a mean life of τ seconds, τ' days, τ'' years; with a half-life of T seconds, T' days, T'' years.

6.10 *

Consider the decay chain

$$X_1 \xrightarrow{\tau_1} X_2 \xrightarrow{\tau_2} X_3 \xrightarrow{\tau_3} X_4 \text{ (stable)},$$

where the mean life of the decay τ_i is indicated over the arrow. Let $N_i(t)$ be the number of nuclei of type X_i present at time t. Derive the coupled differential equations satisfied by the N_i, $i = 1,4$. Obtain an approximate solution for the case $\tau_1 \ll \tau_2 \ll \tau_3$; $N_i(0) = 0$, $i \neq 1$; $N_1(0) = N_0$. Sketch $N_i(t)$ vs. t for the four substances. Sketch the activity of each substance (number of curies) vs. t. Describe what is happening.

6.11 *

Derive Equation 6.6. Obtain an expression for B in terms of known constants and $N_x(E)$, the number of ion pairs created when a particle of type x ($x = \alpha$, β, γ, etc.) and energy E traverses 1 cm of dry air.

How is the derivation modified so as to give the dose in rad rather than in R? What quantity $N'_x(E)$ takes the place of $N_x(E)$?

In traversing a thickness of 1 gram/cm^2 of water, a 1-MeV γ-ray deposits on the average 0.03 MeV of energy. Using this information, calculate N'_γ (1 MeV) and estimate N_γ(1 MeV). With these values of N' and N, find the dose (in both rad and R) received in one hour at a distance of 5 meters from a radioactive source of 1 curie strength, if the decay consists of emission of a 1-MeV γ-ray, and the half-life is long compared to 1 hour.

6.12

When a mammal receives a radiation dose of 1 rad, roughly how many ion pairs are created in a typical cell (diameter about 20μ)? How many are created in a typical cell nucleus where the genetic information is contained (diameter about 6μ)?

6.13

Calculate what fraction of slow-neutron-induced ^{235}U fissions produce ^{137}Cs as a fission product. Use the information given in the third and fourth paragraphs of Section 6.13. (The half-life of ^{137}Cs is 30 years.)

6.14

Convert the radioactive releases listed in Table 6.7 from curies (decays/sec) to total decays by using the following assumptions about half-lives. (1) Liquid releases exclusive of tritium have an effective half-life on the order of 3 years. (2) The half-life of tritium is 12 years. (3) Under good operating conditions (low releases), the gaseous releases from a PWR are dominantly ^{85}Kr (11-year half-life), but when gaseous releases are large, short half-life species predominate and the effective half-life is on the order of 3 days. (4) Gaseous releases from a BWR have effective half-lives dropping from 3 days to 0.5 day as the release increases. (It must be stressed that the effective half-lives given above, involving poorly known mixes of radionuclides, are rough order-of-magnitude estimates only, and your answers will therefore be only order-of-magnitude estimates.)

6.15

Imagine a PWR that continuously releases tritium (12-year half-life) to the environment at a rate of 1000 Ci/yr, and a BWR that continuously releases ^{133}Xe (5-day half-life) to the environment at a rate of 100,000 Ci/yr. For each case, calculate the number of curies present in the environment as a function of time after reactor turn-on. What is the equilibrium value in each case? What is the probable geographical distribution of the radioisotope in each case? Comment on possible harmful results from each situation.

6.16

The only long-lived noble gas radioisotope produced in any quantity in the slow-neutron-induced fission of ^{235}U is ^{85}Kr. It is produced at a rate of 125 Ci/MW$_e$-yr. Because krypton is a chemically inert gas, it is difficult (though not impossible) to retain ^{85}Kr in fuel reprocessing and to store it *in perpetuity*. Current practice is to release it to the atmosphere during fuel reprocessing. Krypton-85 has an 11-year half-life and releases about 1 MeV on decay.

a) Suppose the world nuclear power industry grows in the future to a power level of 6×10^6 MW$_e$ (the present world power level from all sources) and stabilizes at that value. How many curies of ^{85}Kr would be produced each year? Suppose the practice of releasing ^{85}Kr to the atmosphere is continued. What would the equilibrium level of ^{85}Kr be in the atmosphere?

b) Assuming the ^{85}Kr is uniformly distributed throughout the atmosphere, one can calculate the annual dose (in rads) that it gives by dividing the annual energy release from ^{85}Kr-decay by the mass of the atmosphere. (Justify this procedure.) For the equilibrium level found in part (a) above, what is the annual dose? How does this compare with background radiation?

6.17

Small spherical bodies moving through a viscous fluid experience a retarding viscous force given by

$$f = 6 \pi \eta a v,$$

where a is the radius of the sphere, v is its velocity, and η is the viscosity of the fluid.

a) Derive an equation for the terminal velocity of a small spherical particle settling through a viscous fluid under the influence of gravity by balancing the gravitational and viscous forces. (This equation is known as Stokes' Law.)

b) The viscosity of air is 1.8×10^{-5} newton-sec/m^2. Calculate the thermal velocity in air of a 1-μ-diameter particle of density 2g/cm^3. Particles with terminal velocities in the km/hr range (and higher) are quickly removed from the atmosphere (in minutes to hours) because there are no consistent vertical air currents to counteract this rapid downward drift. What diameter particles are in this category? Particles with terminal velocities below the meter/hr range are not removed by gravity to any appreciable extent because gentle vertical air currents overwhelm the effect of gravity. What diameter particles are in this category?

c) Consider a situation in which particulates of a broad range of sizes are carried to a height of 100 meters by the plume from a smokestack, and then move under the influence of the prevailing wind, which is a light breeze of 2 km/hr. There are no vertical air currents. Plot as a function of particle diameter the time a particle will stay aloft and the distance it will travel from the smokestack.

Notes

1

A. Leopold, *A Sand County Almanac, With Other Essays on Conservation from Round River*. New York: Oxford University Press, 1966, p. ix. By permission.

2

Ibid., p. x. By permission.

3

J. B. Wiesner and H. F. York, "National Security and the Nuclear-Test Ban." *Scientific American* **211,** No. 4, p. 27.

4

Garrett Hardin, "The Tragedy of the Commons." *Science* **162,** p. 1243.

5

CEQ, *Environmental Quality,* First Annual Report of the Council on Environmental Quality, Washington, D.C.: U.S. Government Printing Office, 1970, and subsequent annual reports. These annual reports contain a good "official" view of the state of the environment and of important happenings, new legislation, etc. for the year under consideration.

6

SMIC, *Inadvertent Climate Modification,* Report of the Study of Man's Impact on Climate. Cambridge, Mass.: MIT Press, 1971, p. 112.

7

Clear-water ecosystems contain two quite different phases with a sharp boundary between them. There is a phase of "oxygen-rich, nutrient-poor" clear water right up to the edge of a phase of "mineral-rich, oxygen-poor" organic matter (either living or in the form of organic deposits). In the case of freshwater lakes and rivers, the organic phase is beyond the edge of the clear water in the adjacent marshes and shoreline woods. The lake or river bottom is a clean mineral surface nearly depleted of organic matter (and oxidized), because the marsh/woodland acts as a filter and ion exchanger for the inflow and retains reducing organic matter, trace elements, and available phosphate and nitrate. Under the conditions just described, biological activity in the clear water is limited by the supply of nutrients, and the supply of oxygen is ample. Many human activities can break the boundary around the clear-water phase. Dumping sewage effluent into a lake adds organic matter and nutrients; fertilizer runoff from adjacent farmlands adds nutrients; filling marshes or clearing woodlands reduces the area of the "mineral-rich, oxygen-poor" phase and decreases its effectiveness as a filter, allowing more natural organic matter and nutrients to enter the clear-water phase. With the boundary around the clear-water phase not fully effective, the increased concentration of nutrients allows more plant growth, allowing in turn more animal growth. This organic matter, along with the organic matter now entering the clear-water phase, is ultimately oxidized either in respiration or in decay, and oxygen from the clear-water phase is consumed in the process. As a result, the oxygen supply is no longer ample.

8

JCAE, *Selected Materials on Environmental Effects of Producing Electric Power,* U.S. Congress Joint Committee on Atomic Energy. Washington, D.C.: U.S. Government Printing Office, 1969, p. 325.

9

J. G. Asbury, "Effects of Thermal Discharges on the Mass/Energy Balance of Lake Michigan." Report ANL/ES–1, Center for Environmental Studies, Argonne National Laboratory, July 1970.

10

The "official party line" is presented by C. K. Leeper, "How Safe Are Reactor Emergency Cooling Systems?" *Physics Today* **26,** No. 8, p. 30. The view of the "loyal opposition" is presented by Daniel Ford and Henry Kendall, *An Evaluation of Nuclear Reactor Safety,* Union of Concerned Scientists, Cambridge, Mass., 1972; Ford and Kendall, "Nuclear Safety," *Environment* **14,** No. 7, p. 2.

11

J. O. Blomeke, J. P. Nichols, and W. C. McClain, "Managing Radioactive Wastes." *Physics Today* **26,** No. 8, p. 36.

12

EPA studies quoted in CEQ Second Annual Report, August 1971, pp. 104–107.

13

H. C. Hottel and J. B. Howard, *New Energy Technology—Some Facts and Assessments.* Cambridge, Mass.: MIT Press, 1972, pp. 15–16.

14

Ibid., pp. 67–68.

15

E. S. Deevey, "Mineral Cycles." *Scientific American* **223,** No. 3, p. 148.

16

ACS, *Cleaning Our Environment—The Chemical Basis for Action,* American Chemical Society, Washington, D.C., 1968, p. 32. Estimates 1/2. Erik Erikson, *Power Generation and Environmental Change*, D. A. Berkowitz and A. M. Squires (eds.). Cambridge, Mass.: MIT Press, 1971, p. 290. Estimates 1/4.

17

Erik Erikson, p. 299.

18

SMIC, p. 235.

19

Reported in SMIC, p. 238.

20

SCEP, *Man's Impact on the Global Environment*, Report of the Study of Critical Environmental Problems. Cambridge, Mass.: MIT Press, 1970, p. 161.

21

Harry Perry, "Environmental Aspects of Coal Mining," in *Power Generation and Environmental Change*, D. A. Berkowitz and A. M. Squires (eds.). Cambridge, Mass.: MIT Press, 1971, p. 331.

22

Gerald Garvey, *Energy, Ecology, Economy*. New York: W. W. Norton, 1972, p. 83.

23

Ibid., p. 88.

24

SCEP, pp. 142–143.

25

Final Environmental Statement for the Prototype Oil Shale Leasing Program, Vols. 1–6. U.S. Department of the Interior, 1973.

26

See, for example, *Engineering for Resolution of the Energy-Environment Dilemma*. Washington, D.C.: National Academy of Engineering, 1972, p. 100.

Related Reading

The reader with no prior background in environmental problems will find it advantageous to read one or more of the following four books.

A. Turk, J. Turk, and J. T. Wittes
Ecology, Pollution, Environment. Philadelphia: W. B. Saunders, 1972.

Raymond F. Dasmann
Environmental Conservation, 3rd edition. New York: Wiley, 1972.

Richard H. Wagner
Environment and Man, 2nd edition. New York: W. W. Norton, 1974.

William W. Murdock (ed).
Environment, Resources, Pollution, and Society. Stamford, Conn.: Sinauer Associates, 1971.

The following two books discuss in economic terms the inadequacy of the free enterprise system in coping with environmental problems.

T. D. Crocker and A. J. Rogers III
Environmental Economics. Hinsdale, Illinois: Dryden Press, 1971.

Paul W. Barkley and David W. Seckler
Economic Growth and Environmental Decay. New York: Harcourt, Brace, Jovanovich, 1972.

The remaining references contain a wealth of information and opinion on energy-related environmental problems. Fact and opinion are not always identified as such; watch out for author's bias!

General References

OST
Electric Power and Environment, Energy Policy Staff, Office of Science and Technology. Washington, D.C.: U.S. Government Printing Office, 1970.

Environmental Effects of Producing Electric Power, Parts I and II. Hearings before the Joint Committee on Atomic Energy, U.S. Congress. Washington, D.C.: U.S. Government Printing Office, 1969 and 1970.

JCAE
Selected Materials on Environmental Effects of Producing Electric Power, Joint Committee on Atomic Energy, U.S. Congress. Washington, D.C.: U.S. Government Printing Office, 1969.

JEC
The Economy, Energy and the Environment, Background study prepared for the Joint Economic Committee, 91st U.S. Congress. Washington, D.C.: U.S. Government Printing Office, 1970.

Summary Report of the Cornell Workshop on Energy and the Environment, February 22–24, 1972. Washington, D.C.: U.S. Government Printing Office, 1972.

CEQ
Energy and the Environment—Electric Power, Report prepared by the Council on Environmental Quality. Washington, D.C.: U.S. Government Printing Office, 1973.

Dean E. Abrahamson
Environmental Cost of Electric Power, A SIPI Workbook. Scientist's Institute for Public Information, New York, 1970.

APHA
Health Effects of Energy Systems—A Quantitative Assessment, Report of the American Public Health Association Task Force on Energy to the Ford Foundation Energy Policy Project, submitted March 1974. Compares the direct public health effects of various different sources of energy on the basis of deaths or injuries per 10^{15} Btu.

NAE
Engineering for Resolution of the Energy-Environment Dilemma, Report of the Committee on Power Plant Siting. Washington, D.C.: National Academy of Engineering, 1972.

Gerald Garvey
Energy, Ecology, Economy. New York: W. W. Norton, 1972. A survey of energy's environmental problems from an economist's point of view.

John Holdren and Philip Herrera
Energy, A Crisis in Power. San Francisco: Sierra Club, 1971. A broad, quick survey

of the environmental problems associated with energy, followed by case studies of some famous recent controversies.

David A. Berkowitz and Arthur M. Squires (eds.)
Power Generation and Environmental Change, Symposium of the Committee on Environmental Alteration, American Association for the Advancement of Science, December 28, 1969. Cambridge, Mass.: MIT Press, 1971. Contains several different viewpoints on radiation dangers. Also has articles on coal mining and coal combustion, hydroelectric power, air pollution, and thermal pollution.

Thermal Pollution

Theodore L. Brown
Energy and the Environment, Columbus, Ohio: Charles E. Merrill, 1971. A treatment of thermal pollution, greenhouse effects, and inadvertent climate modification. Aimed at the general public.

SMIC
Inadvertent Climate Modification, Report of the Study of Man's Impact on Climate. Cambridge, Mass.: MIT Press, 1971. A consideration of large-scale thermal effects.

John R. Clark
"Thermal Pollution and Aquatic Life," *Scientific American* **220,** No. 3, p. 18.

Thermal Pollution, 1968, Hearings before the Subcommittee on Air and Water Pollution of the Committee on Public Works, U.S. Senate, 90th Congress. Washington, D.C.: U.S. Government Printing Office, 1968.

Radioactivity

David R. Inglis
Nuclear Energy: Its Physics and Its Social Challenge. Reading, Mass.: Addison-Wesley, 1973. Aimed at the general public.

John W. Gofman and Arthur R. Tamplin
Poisoned Power—The Case Against Nuclear Power Plants. Emmaus, Pa.: Rodale Press, 1971. Two scientists with considerable training in the field present their case against using nuclear power. The book is written for a general audience. Some rebuttals to Gofman and Tamplin appear in Berkowitz and Squires (eds.).

AEC
Environmental Survey of the Nuclear Fuel Cycle, Report prepared by the U.S. Atomic Energy Commission, Washington, D.C., 1972. Considers all aspects of the LWR fuel cycle except the nuclear reactor. This document is the AEC "party line," sweeps many past accidents under the rug, and should be read with a critical eye. The treatment of storage of high-level wastes in a repository yet to be designated is particularly unsatisfactory.

Daniel Ford, Thomas Hollocher, Henry Kendall, et al.
The Nuclear Fuel Cycle. San Francisco: Friends of the Earth, 1974. A critique of the AEC document by the "loyal opposition."

Air Pollution

Samuel J. Williamson
Fundamentals of Air Pollution. Reading, Mass.: Addison-Wesley, 1973. An introductory text on the subject. Assumes the reader has a scientific background.

Arthur C. Stern (ed.)
Air Pollution, Vols. 1, 2, and 3. New York: Academic Press, 1968. This three-volume compendium is aimed at the expert and contains much useful information.

Hans A. Panofsky
"Air Pollution Meteorology," *American Scientist* **57,** pp. 269–285. A nice treatment of the local movement of air pollutants.

For climate modification aspects of air pollution, also see T. L. Brown and SMIC under "thermal pollution."

7 Energy Use: History and Projections

7.1 Energy Use versus Time

From its earliest days, the species *Homo sapiens* has used energy. Initially man used only food energy, as did all other species of animals. Cultural development led to the use of other energy sources and also to increases in human population. It is only comparatively recently that records of world population have been kept, and even more recently that records of world energy use have been kept. For the long view of energy use, we must rely on estimates rather than direct records. Table 7.1 gives estimates of world population at various times, the per capita energy use for the most advanced culture at that time, and the world average per capita energy use. Primitive man's use of 2000 kcal/day was entirely food energy. Hunting man used some fuel wood for heat and cooking, and early agricultural man also used animal energy for agricultural and transportation. Advanced agricultural man added water and wind power.

The "natural level" of human energy use can be taken as that of primitive man, namely, 2000 kcal/man-day times one million people, or 2×10^9 kcal/day. Current human energy use is 60,000 kcal/man-day times four billion people, or 2.4×10^{14} kcal/day. This increase by a factor of 10^5 is attributable largely to population growth, and to a much smaller extent to increased per capita energy use. (Of course, the large population growth was made possible in part because of increased per capita energy use.) Since 1400 A.D., the two factors have been more nearly comparable in their importance.

The per capita energy use for the United States, India, and the entire world for the past hundred years or so is shown in Fig. 7.2. The current world per

187

Table 7.1 Growth in population and per capita energy use. Estimates of energy use for the most advanced culture are by E. Cook. World average energy estimates were inferred from these.

Year	World Population	World average per capita energy use (kcal/man-day)	Most advanced culture	Per capita energy use, advanced culture (kcal/man-day)
10^6 years ago	1×10^6	2×10^3	Primitive man	2×10^3
10^5 years ago	2×10^6	3×10^3	Hunting man (Europe)	5×10^3
5000 B.C.	35×10^6	9×10^3	Early agricultural man (Fertile Crescent)	12×10^3
1400 A.D.	0.5×10^9	15×10^3	Advanced agricultural man (Northwestern Europe)	26×10^3
1875	1.5×10^9	25×10^3	Industrial man (England)	77×10^3
1970	4×10^9	60×10^3	Technological man (United States)	230×10^3

Source: Adapted from E. Cook, "The Flow of Energy in an Industrial Society," *Sci. Amer.* **224,** p. 136. Copyright © 1971 by Scientific American, Inc. All rights reserved.

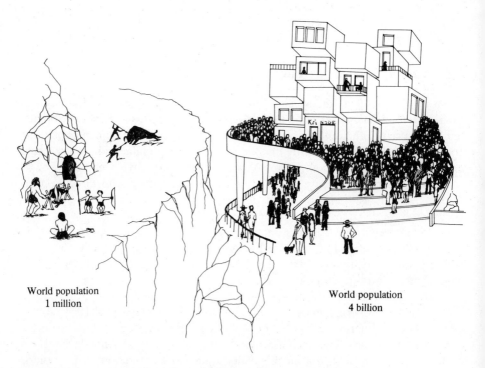

World population
1 million

World population
4 billion

Fig. 7.1(a) Population then and now. (Sketch courtesy of H. Soosaar.)

2,000 Calories   230,000 Calories

Fig. 7.1(b) Energy use then and now. (Sketch courtesy of H. Soosaar.)

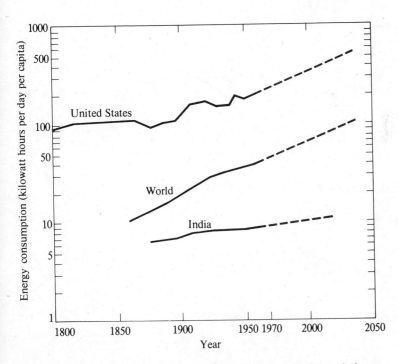

Fig. 7.2 Per capita energy use in the United States, India, and the entire world in recent decades, with projections beyond the year 2000. (From Chauncey Starr, "Energy and Power," *Sci. Amer.* **225**, p. 40. Copyright © 1971 by Scientific American, Inc. All rights reserved.)

capita energy use is about one-fifth that of the United States. If one subtracts the energy consumption of the United States and the other industrialized nations, then the rest of the world, which represents well over half of the world's population, is using energy at a per-capita level less than one-tenth of the United States. India, at one-twenty-fifth the U.S. per capita energy level, is illustrative. It is probable that the figure does not fully include such nonindustrial sources of energy as food, animal power, and cow dung; inclusion of these would raise the curve for India and for the nonindustrialized world. Nevertheless, a substantial fraction of the world's population is using energy at or below the level of the "advanced agricultural man" of 1400 A.D. Europe.

Per capita energy use in the United States has always been high, even as far back as 1800 when the United States was primarily a rural agricultural nation. This anomaly can be traced to the country's relatively recent development. Prior to 1600, the human population in North America was minimal, and as a result the extensive forests remained intact. These forests furnished an inexpensive, readily available energy supply nearly until the end of the nineteenth century. By contrast, the forests of Europe and Asia were extensively cut much earlier, and fuel wood was therefore less abundant.

In addition to changes in the magnitude of energy use, we must consider

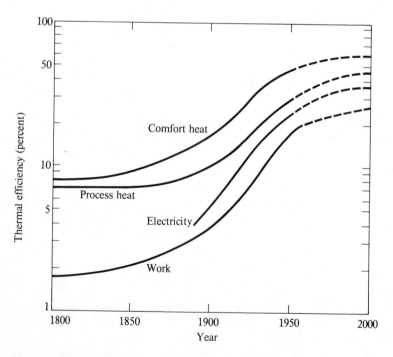

Fig. 7.3 Efficiencies of various energy converters since the year 1800, with projections to the year 2000. (From Chauncey Starr, "Energy and Power," *Sci. Amer.* **225,** p. 40. Copyright © 1971 by Scientific American, Inc. All rights reserved.)

changes in the efficiency of energy use. This comparison is shown in Fig. 7.3. From 1800 to 1900, efficiencies approximately doubled. During this same period, U.S. per capita energy use remained constant, so useful energy per capita doubled. From 1900 to 1950, efficiencies increased by a factor of 4 and per capita energy use doubled. Therefore, useful energy per capita went up by a factor of 8. This is a very large increase for a 50-year period. In the future, efficiencies will probably not increase appreciably, certainly not as dramatically as during the first 50 years of the twentieth century. Increases in useful energy must come dominantly from increases in energy use.

Sources of energy have changed with time, as is shown in Fig. 7.4. Fuel wood was the dominant U.S. energy source until 1880, at which time coal took over. By 1940, coal lost out to oil and natural gas. The expectation is that soon uranium will be supplying substantial amounts of energy. In addition to the shifts among the primary energy sources, there has been a shift from direct use of fuel to use of electricity generated with the fuel. This shift has been very pronounced ever since commercial electrical power was introduced in 1882. Electrical energy use has doubled about every 10 years since then, while total energy use has doubled about every 20 years. Currently 25 percent of the primary energy is used for generation of electricity; by 2000 A.D. it is projected that this figure will rise to 50 percent.

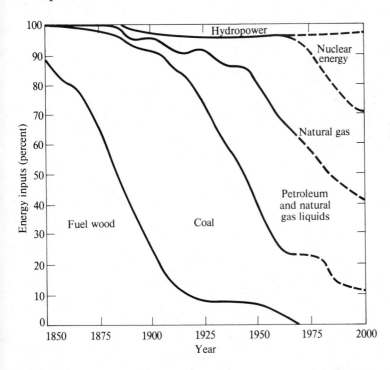

Fig. 7.4 Sources of U.S. energy *vs.* time from the year 1850, with projections to the year 2000. (From Earl Cook, "The Flow of Energy in an Industrial Society," *Sci. Amer.* **225**, p. 137. Copyright © 1971 by Scientific American, Inc. All rights reserved.)

7.2   Uses of Energy

The end uses of energy for the various sectors are listed in Table 7.2. The industrial sector is broken down into standard industrial classifications and individual industries in Table 7.3. The transportation sector is broken down into modes of

Table 7.2   Energy consumption in the United States by end use, 1960 to 1968.

Sector and end use	Consumption (10¹² Btu/yr) 1960	1968	Annual rate of growth (percent/yr)	Percent of national total 1960	1968
Residential					
Space heating	4,848	6,675	4.1	11.3	11.0
Water heating	1,159	1,736	5.2	2.7	2.9
Cooking	556	637	1.7	1.3	1.1
Clothes drying	93	208	10.6	0.2	0.3
Refrigeration	369	692	8.2	0.9	1.1
Air conditioning	134	427	15.6	0.3	0.7
Other	809	1,241	5.5	1.9	2.1
Total	7,968	11,616	4.8	18.6	19.2
(electrical component)	(2,054)	(3,818)	(8.2)	(4.8)	(6.3)
Commercial					
Space heating	3,111	4,182	3.8	7.2	6.9
Water heating	544	653	2.3	1.3	1.1
Cooking	98	139	4.5	0.2	0.2
Refrigeration	534	670	2.9	1.2	1.1
Air conditioning	576	1,113	8.6	1.3	1.8
Feedstock	734	984	3.7	1.7	1.6
Other	145	1,025	28.0	0.3	1.7
Total	5,742	8,766	5.4	13.2	14.4
(electrical component)	(1,439)	(2,964)	(9.6)	(3.3)	(4.9)
Industrial					
Process steam	7,646	10,132	3.6	17.8	16.7
Electric drive	3,170	4,794	5.3	7.4	7.9
Electrolytic processes	486	705	4.8	1.1	1.2
Direct heat	5,550	6,929	2.8	12.9	11.5
Feedstock	1,370	2,202	6.1	3.2	3.6
Other	118	198	6.7	0.3	0.3
Total	18,340	24,960	3.9	42.7	41.2
(electrical component)	(3,615)	(5,612)	(5.8)	(8.4)	(9.3)
Transportation					
Fuel	10,873	15,038	4.1	25.2	24.9
Raw materials	141	146	0.4	0.3	0.3
Total	11,014	15,184	4.1	25.5	25.2
(electrical component)	(50)	(49)	—	(0.1)	(0.1)
National total	43,064	60,526	4.3%	100.0%	100.0%
(electrical component)	(7,159)	(12,443)	(7.1%)	(16.6%)	(20.6%)

Note: Electric utility consumption is allocated to each end use and includes generating losses.

Source: *Patterns of Energy Consumption in the United States.* Office of Science and Technology, Executive Office of the President, Washington, D.C., January 1972.

Table 7.3   Percentage of 1968 industrial energy use for various industrial classifications and specific industries.

Standard industrial classification	
Primary metal industries	21.2%
Chemicals and allied products	19.8
Petroleum refining and related industries	11.3
Food and kindred products	5.3
Paper and allied products	5.2
Stone, clay, glass, and concrete products	4.9
All other industries	32.3
	100.0%

Specific industries	
Iron and steel	13.6%
Petroleum refining	11.3
Paper and paperboard	5.2
Petrochemical feedstock	4.9
Aluminum	2.8
Cement	2.1
Ammonia	2.0
Ferrous foundries	2.0
Carbon black	0.9
Grain mills	0.8
Copper	0.8
Glass	0.8
Concrete	0.7
Meat products	0.7
Soda ash	0.7
Sugar	0.7
All others	50.0
	100.0%

Source: *Patterns of Energy Consumption in the United States.* Office of Science and Technology, Executive Office of the President, Washington, D.C., January 1972.

transportation in Table 7.4. The contributions of the primary energy sources to the various use sectors are given in Table 7.5. You should study these tables and play with them. (Problems 7.4 through 7.7 will give you the opportunity.)

From a perusal of these tables, we see that lots of things contribute to energy use. No one sector or end use dominates. The largest sector, industrial, is composed of many individual industries, none of which constitutes as much as 6 percent of national energy use. Two end uses do attract attention: Residential and commercial space heating account for 18 percent of all energy use, and automobile use accounts for another 14 percent. Concerning sources of energy, we note that coal is used mainly for generation of electricity and for industrial purposes, and that the transportation sector is supplied almost exclusively by petroleum.

A word of warning should be given about all of these tables. The precision of the number is considerably less than indicated by the number of significant figures given. In particular, in Tables 7.2 and 7.4, the "other" subcategories are

Table 7.4 Distribution of energy use within the transportation sector.

Mode of transportation	Percent of total transportation energy	
	1960	1970
Automobiles		
Urban	25.2	28.9
Intercity	27.6	26.4
	(52.8)	(55.3)
Aircraft		
Freight	0.3	0.8
Passenger	3.8	6.7
	(4.1)	(7.5)
Railroads		
Freight	3.7	3.2
Passenger	0.3	0.1
	(4.0)	(3.3)
Trucks		
Intercity freight	6.1	5.8
Other uses	13.8	15.3
	(19.9)	(21.1)
Waterways, freight	1.1	1.0
Pipelines	0.9	1.2
Buses	0.2	0.2
Other[a]	17.0	10.4
Total	100.0%	100.0%
Total transportation energy consumption	10.9×10^{15} Btu	16.5×10^{15} Btu

[a] Includes passenger traffic by boat, general aviation, pleasure boating, and nonbus urban mass transit, as well as the effects of historical variations in modal energy efficiencies.

Source: Eric Hirst, *Energy Consumption for Transportation in the U.S.*, ORNL-NSF-EP-15, March 1972.

obtained by subtracting the estimates for all the listed categories from the total, which therefore compounds the errors from all these estimates.

A comparison of U.S. energy use with that of other industrialized countries is interesting. Figures for Great Britain are given in Table 7.6. (The "other" category of Table 7.6 is roughly equivalent to the "commercial" category of Table 7.2, though some commercial uses probably appear in the "industry" category of Table 7.6.) During 1971, Great Britain had slightly more than half the per capita GNP and slightly less than half the per capita energy use of the United States. The price of energy was about twice as high there as here.

Comparison of Tables 7.2 and 7.6 reveals that the major change in distribution among the end-use sectors was the increased percentage for domestic uses and the decreased percentage for transportation in Great Britain. On a per-capita basis, British energy use for domestic, industrial and other, and transportation were 66 percent, 44 percent, and 27 percent, respectively, of U.S. per capita energy use. The difference in transportation is striking.

Table 7.5   Contribution of energy sources to use sectors. Electricity appears both as a use sector and as an energy source. As a source, it is listed both including and excluding generating losses. Average generating efficiency is 36.4 percent. Although hydropower supplies only 6 percent of the energy for electric generation, it supplies 16 percent of the generated electricity because of its 100 percent generating efficiency. Figures are trillion Btu's per year for the year 1968.

Energy source	Residential	Commercial	Transportation	Industrial	Electric generation	Total
		4.4%	0.1%	42%	53%	100%
Coal	—	586	12	5,616	7,130	13,326
					57%	
	12%	13%	54%	17%	4%	100%
Oil	3,192	3,389	14,513	4,474	1,181	26,749
					10%	
	24%	9%	3%	47%	17%	100%
Natural gas	4,606	1,845	610	9,258	3,245	19,564
					26%	
					100%	100%
Hydropower	—	—	—	—	757	757
					6%	
					100%	100%
Nuclear energy	—	—	—	—	130	130
					1%	
	31%	24%	0.4%	45%		100%
Electricity (including	1,390	1,079	18	2,043	—	4,530
generating losses)	(3,818)	(2,964)	(49)	(5,612)	—	(12,443)
Total	11,616	8,766	15,194	24,960	12,443	**60,526**
					100%	

Source: *Patterns of Energy Consumption in the United States.* Office of Science and Technology, Executive Office of the President, Washington, D.C., January 1972.

Table 7.6   Energy consumption in Great Britain by end uses for 1971. The population was 56 million and the annual GNP was $150 billion for that year.

	Direct use (10^{14} Btu)	*Via* electricity (10^{14} Btu)	Total (10^{14} Btu)	Percent
Total	57.0	18.7	75.7	100
Domestic	14.2	7.4	21.6	28.5
Iron and steel	6.6	1.0	7.6	10
Other industry	17.5	6.4	23.9	31.5
Transportation	11.6	—	11.6	15
Other	7.2	3.9	11.1	15

Source: "United Kingdom Energy Statistics," Government Statistical Service, Department of Trade and Industry, United Kingdom, 1973.

7.3   Some Economic Considerations

Price

The price of energy varies considerably with the particular circumstances of its extraction and transportation, with the quantity of the purchase, and with time. For these reasons the numbers given below should not be taken too literally. (Most transactions during the early 1970's would have differed from the numbers quoted by less than a factor of 2. Since the winter of 1973–1974, prices, particularly those for oil, have been in a state of flux and are still changing too rapidly to allow any statements about "present" prices to be made.)

A typical "mine-mouth" price for eastern coal is $4 per ton; a typical delivered price paid by a large purchaser (e.g., a utility company) is $8 per ton. These prices correspond to $0.15/10^6$ Btu and $0.30/10^6$ Btu, respectively.

The situation with oil is complicated. Partly because of land ownership patterns and partly because of geology, oil in the United States is produced from a large number of wells, each yielding a small amount of oil. However, oil in many foreign countries, particularly in the Middle East, comes from a small number of wells, each yielding a large amount of oil. For this reason production costs in the Middle East are considerably below those in the United States. Finding costs are also lower in the Middle East. With the advent of large tankers, transportation costs became sufficiently small that foreign oil could undersell U.S. domestic oil, and import quotas were established to protect domestic producers. However, since 1950 U.S. domestic production has been unable to meet demand, and imports, principally from Canada and Venezuela, have been growing. During the 1970's, as demand for oil continued to increase (in the United States, Europe, and Japan) and as Arab politicians more fully appreciated the value of their export, the oil producing and exporting countries (OPEC) collectively decided to raise their prices. Following the 1973 Arab-Israeli war, the OPEC reduced production and established boycotts on some countries. All these steps resulted in a short supply of oil and prices that rapidly rose. During the early 1970's, $3 to $4/bbl for crude oil delivered to a refinery was the price for domestic crude, and imported crude was cheaper. By 1974, imported crude sold for $12/bbl, and Congress considered limiting the price of domestic crude to not more than $7/bbl. ($4/bbl corresponds to $0.70/10^6$ Btu; $7/bbl to $1.20/10^6$ Btu.) A typical price in 1972 for home heating oil (for a residential customer) was $0.20/gal ($1.40/10^6$ Btu); the price of gasoline (with its higher retailing costs and including taxes) was in the neighborhood of $0.40/gal. In the fall of 1974, the price of home heating oil was $0.37/gal and gasoline was $0.60/gal.

Before pipeline shipment, natural gas costs about $0.30/1000 cu ft; in an eastern U.S. city, the price to a large industrial user is $0.65/1000 cu ft. Delivered to a residential customer, it sells for $1.65/1000 cu ft. These prices correspond to $0.28/10^6$ Btu, $0.62/10^6$ Btu, and $1.54/10^6$ Btu, respectively.

The cost of generating (but not distributing) electricity is 7 to 10 mils/kWh. The price to a residential customer is $0.02 to $0.03/kWh; the price to a large industrial customer can be quite close to the generating cost.

Both electricity and natural gas are sold according to a decreasing rate structure, by which the cost of an incremental unit of energy decreases as the total amount of energy purchased per month increases. That is, the last kilowatt hour of electricity or cubic foot of natural gas in a given month costs less than the first kilowatt hour or cubic foot. An electricity rate schedule for a residential user is shown in Table 7.7. The justification of a decreasing rate structure and the arguments against it are presented in Section 9.1.

Table 7.7   Electricity rate schedule for residential service in 1974.

Electricity used in a month	Cost
First 12 kWh, or less	$1.75
Next 38 kWh, per kWh	$0.0519
Next 50 kWh, per kWh	$0.0395
Next 200 kWh, per kWh	$0.0282
Next 500 kWh, per kWh	$0.0218
Next 1200 kWh, per kWh	$0.0197
Over 2000 kWh, per kWh	$0.0184

Source: Rochester Gas and Electric Corporation, August 1974.

Synthetic oil and gas can be made from coal. A 1971 estimate[1] for the wholesale price of gas from coal is $1/1000 cu ft; that of oil from coal is $4/bbl. More recent estimates are higher. The fact that foreign oil *can* sell for considerably less than $4/bbl has inhibited the research and development work and the capital commitment necessary for a large-scale oil-from-coal plant. The existence of such a plant would provide some protection against escalating foreign-oil prices.

Relation of Energy to GNP

From the preceding, $1/10^6$ Btu is a reasonable figure for the price of energy in the early 1970's, averaging over industrial and residential uses. With an annual energy use of 65×10^{15} Btu (appropriate for 1970), the annual U.S. energy bill was $65 billion. This is small compared to the annual GNP of about $1000 billion. (More careful studies suggest that the energy industries constitute about 5 percent of the GNP, confirming our simple estimate.)

The cost of energy as a percentage of the value of goods produced is shown for several industries in Table 7.8. There is great variability from industry to industry, but on the average the dollar contribution of energy to total product price is quite small.

The low dollar contribution of energy to GNP should not be interpreted as meaning that energy plays no important role in producing GNP. Clearly, most of the components of GNP require significant amounts of energy for their production. The essential nature of energy is shown in Fig. 7.5, which plots per capita

Table 7.8 Value of fuel and electricity purchases as a percentage of value of goods produced, for various industries.

Industry	Energy costs / Value of goods produced
All manufacturing	1.4%
Primary nonferrous metals	5.5%
Hydraulic cement	15.4%
Basic chemical	6.0%
Steel rolling and finishing (including blast furnaces)	4.1%
Machinery	0.65%
Printing and publishing	0.5%

Source: Adapted from S. H. Schurr (ed.), *Energy, Economic Growth, and the Environment.* Published for Resources for the Future, Inc. Baltimore: The Johns Hopkins University Press, 1972, p. 157. By permission.

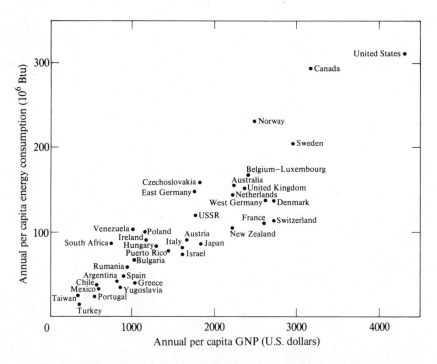

Fig. 7.5 Annual per capita energy consumption *vs.* annual per capita GNP for several countries for 1968. (Data from S. H. Schurr (ed.), *Energy, Economic Growth, and the Environment.* Published for Resources for the Future, Inc. Baltimore: The Johns Hopkins University Press, 1972, pp. 182–183. By permission.)

GNP against per capita energy use for several countries. The correlation, though not perfect, is very strong. The deviations from a perfect correlation can be understood partly in terms of the nature of a country's output. Those countries stressing agriculture require somewhat less energy than the average to produce a given amount of GNP. Those countries stressing heavy industry require somewhat more energy than the average. Also, the availability of energy (i.e., price) will have an influence. Where energy is abundant and inexpensive, it will be used less efficiently than where it is scarce. Finally, in the colder climates more energy is used for space heating. The deviations should not obscure the inescapable conclusion that it takes energy to produce GNP.

Imports

Until 1950, the United States was a net exporter of energy, exporting coal in significant quantities. Since that time, U.S. oil imports have risen rapidly, and more recently, imports of natural gas have become significant. In 1970, the United States imported 12 percent of its energy needs (mainly in the form of oil), while exporting only one-quarter of this amount (in the form of coal). Some projections suggest that by 1980, the United States will be importing one-third of its total energy needs.

A large U.S. import of energy is cause for concern. Because energy plays such a fundamental role in the functioning of our nation, this represents a major lack of self-sufficiency with national security and foreign policy implications. It affects balance of payments significantly. With foreign trade being about 5 percent of GNP ($50 billion in 1972), importing one-third of U.S. energy needs in the form of oil at $7/bbl requires a 50-percent increase in exports to compensate. Finally, the United States is richly endowed in energy resources compared to most other countries and already uses "more than its share" on a per-capita basis. Ethical objections can be raised to the United States importing large amounts of oil.

7.4   Projections Into the Future

There have been a large number of forecasts projecting U.S. energy use in the next few decades.[2] The projected annual energy growth rate ranges from 2.5 percent to 4 percent. A typical study assumes a 4-percent annual GNP increase, a 1.6-percent annual population increase, and projects a 3.2-percent annual energy increase. (For concreteness, we shall define a "standard projection" as a 3.5-percent energy growth rate, implying a doubling time of 20 years.)

U.S. electrical energy use is growing at a rate of 7 percent per year, doubling every 10 years. It has been doing this from the beginning of the twentieth century. About one-quarter of present primary energy sources passes through a stage as electrical energy before final use. Projections suggest that electrical energy growth will remain rapid for a decade or two until electrical energy represents half of total energy. Then electrical energy growth will slow down to that of total energy growth.

World total energy growth slightly exceeds that of the United States, with a growth rate currently over 4 percent. Most projections suggest that this rate will continue. World electrical energy growth is also slightly higher than in the United States, and worldwide electricity generation now uses about one-quarter of the primary energy sources.

All projections are primarily continuations of the trends of the recent past. To understand the reasons for the projected U.S. energy increases, we must examine in more detail the increases that occurred in the recent past. Table 7.2 shows end uses for energy in 1960 and 1968, and lists the average annual increase for each category. The total energy increase was at an average rate of 4.3 percent per year. For the same period, the GNP (corrected for inflation) rose 4.8 percent per year. The relevant population growth figures are not those for the 1960–1968 period, but rather are those for a few decades earlier, since the increase in number of people in the 20-to-40-year age group has a much more substantial impact on energy use than the increase in number of people in the 0-to-10-year age group. A reasonable figure, averaged over the past several decades, is 1.4 percent per year. We see from these figures that per capita energy use is increasing at a rate of 3 percent per year, about twice as fast as its rate in the first half of the twentieth century. Recall, however, the comment in Section 7.1 about efficiency. Useful energy per capita is probably increasing at about the same rate as before, or perhaps even a bit slower.

It is interesting to note that the percentage increase for all four sectors was about the same. Further, most of the large specific end uses increased by a percentage close to the total increase. There were some items that increased by large percentages, but they were items that were small to begin with and therefore did not significantly affect the total growth. (Items growing more rapidly than twice the total growth rate accounted for only 10 percent of the total growth.)

Energy use in the industrial sector increased by 3.9 percent per year, somewhat slower than the GNP increase. That the two increases were not equal can be explained by a shift in GNP from goods toward services and from energy-intensive goods toward less-energy-intensive goods. If GNP per capita is to increase in the future, and unless there is a major shift in the composition of GNP toward services and less-energy-intensive goods, then industrial energy will continue increasing more rapidly than population.

Energy use in the transportation sector increased by 4.1 percent per year, considerably faster than population growth. Since the majority of transportation energy is for passengers rather than freight, this implies that either individuals are traveling more miles per year or they are traveling less efficiently, using more energy per mile. Even though automobile efficiency did fall somewhat between 1960 and 1968, the dominant factor was an increase in miles traveled per individual per year. A further discussion of transportation energy is presented in Chapter 8.

It is informative to try to understand why energy use in the residential sector increased so much faster than population growth. Often this is blamed on "gadgets" (electric toothbrushes, can openers, carving knives, etc.). A look at Table 7.2 shows that this is not so. All small appliances are lumped into the "other" category and collectively account for less than half of it. (These devices generally consume more energy in their manufacture and sales than in their use.)

Residential air conditioning is a very rapidly growing use of energy (15.6 percent per year), but one that is currently very small. By 1968, about 10 percent of all residences had central air conditioning, and perhaps twice as many had room air conditioners. This use has plenty of room for expansion and will probably contribute significantly to future energy growth.

Essentially all households had refrigerators in both 1960 and 1968. The increase in energy use for this category (8 percent per year) greatly exceeded the increase in number of households (2 percent per year), and is due to the increased energy consumption per refrigerator caused by increased size, automatic defrosting, and a larger freezer compartment. Future developments are difficult to predict.

Energy for cooking increased at a rate (1.7 percent) intermediate between the population increase (1.4 percent) and the increase in number of households (2 percent). This is what would be expected for a larger number of households containing on the average fewer members.

The increase in energy for water heating is attributable: (1) in part to the fact that in 1960, 26 percent of all households were without gas or electric water heaters, whereas by 1968, this figure had dropped to 8 percent (a 2.25 percent/yr effect); (2) in part to an increase in the number of households (a 2 percent/yr effect); and (3) in part to increased hot-water consumption (e.g., for dishwashers) (a 0.75 percent/yr effect). For the future, we would expect that the rate of growth in energy for water heating would drop down to the rate of population growth.

Half of the entire increase in energy use in the residential sector is due to space heating. Therefore we should examine that item rather carefully. The increase is 4.1 percent/yr, compared to an increase in number of households of 2 percent/yr, implying that the heating energy per household has been increasing by 2.1 percent/yr. One trend in home heating worth noting is the rapid increase in electrically heated homes. In 1960, 0.7 million homes were heated electrically, whereas by 1968, 3.4 million homes were so heated, giving a rate of increase of 22 percent/yr. A home furnace is typically 75 percent efficient. While conversion of electricity to heat is 100 percent efficient, generation of electricity is only 40 percent efficient, making the overall efficiency of electrical heating about half that of direct use of fuel. Had the incremental $3.4 - 0.7 = 2.8$ million homes been heated with oil or gas rather than electricity, 170 trillion Btu of fuel would have been saved, about 10 percent of the total increase. Thus the trend toward electric heating accounts for a 0.4 percent/yr increase in energy use. (Usually the installation of electric heating is accompanied by better-than-average insulation, probably making 0.4 percent an overestimate.) The remaining 1.7 percent/yr has not been pinpointed but is probably due to a combination of (1) an increase in fraction of single family dwellings at the expense of duplexes and apartments, (2) an increase in the average size of dwellings, (3) ownership of a second home resulting in more than one heated residence per household, and (4) an increase in the temperature level to which dwellings are heated. Geographical shifts of population will also influence home heating requirements, but in recent years the shifts have been toward warmer climates. A further discussion of space heating appears in Chapter 8.

The commercial sector is defined as all activities other than mining and manufacturing, transportation, and residential. It thus contains a very diverse collection of activities, and the figures for the specific end uses given in Table 7.2 were obtained rather indirectly. In particular, the "other" category consists of all electrical energy use not otherwise accounted for. The apparent dramatic increase in this category may in fact be due to an underestimate of the increases due to air conditioning, refrigeration, cooking, water heating, or electrical space heating. In the absence of more reliable end-use figures, no useful comments can be made on how future trends might deviate from those of the recent past.

Electrical energy use has grown faster than total energy use in all sectors except transportation (in which it is negligible in any case). In the case of the residential sector, the end uses causing the growth can be identified with reasonable certainty. The contributions to increased electrical energy use between 1960 and 1968 are electrical space heating (21 percent), water heating (10 percent), cooking and clothes drying (8 percent), refrigeration (20 percent), air conditioning (17 percent), and other (24 percent). Electrical space heating is the only significant case in which the increased electrical energy use is a replacement for direct fuel use. It is electrical space heating, air conditioning, and refrigeration that have caused electrical energy use to grow faster than total energy use in the residential sector. In the commercial sector, air conditioning is responsible for part of the rapid growth of electrical energy use. The rest is hidden in the "other" category.

Increases in electrical energy use in the near future will depend significantly on trends in air conditioning and space heating and could be more or less rapid than in the past.

Questions and Problems

7.1

a) Using Table 7.1, compute the average doubling time for the world human population over the past million years (the duration of *Homo sapiens'* stay on earth). Compute the average doubling time for the world population since civilization began (approximately 5000 B.C.). For the same two time periods, what is the average doubling time for worldwide human energy use? (For comparison, the present-day doubling time for world population is 35 years, and that for world energy use is 20 years.)

b) An estimate of world population a million years ago is fraught with uncertainty, but the world population then was at least 2. Using this much lower value, how do your answers for part (a) change?

c) In order for an individual to *perceive* a change, there has to be a significant change during his or her lifetime. While doubling times for population and energy use were long compared to the human lifespan, the average human would have perceived these quantities as remaining static. Taking (doubling time) = (life span) as an approximation to the perception threshold, when did population increase and energy-use increase first become apparent on a worldwide average basis?

7.2

Using the data given in Table 7.1, display total human power use (in watts) against time before the present (in years) in some graphical manner. (A log-log plot will work well.) Draw a smooth curve through the six points.

a) Integrate human power use over time from the "beginning" to the present to obtain the total amount of energy used to date by the species *Homo sapiens*. What fraction of this energy was used during the first 90 percent of man's stay on earth? What fraction was used during the last 0.1 percent?

b) Energy consumed is a reasonable measure of total human activity during a given period, and is certainly better than time duration, the length of the period in years, and for most purposes better than man-years lived, the product of population during the period times the length of the period. Divide man's stay on earth into ten consecutive "epochs" of equal length in terms of energy consumed and note the length in years of each. Projecting current energy-use figures into the future, how many years will the eleventh "epoch" take?

7.3

From U.S. per capita energy consumption (Fig. 7.2), the distribution of energy inputs among the different sources (Fig. 7.4), and U.S. population figures (not given here, but readily available), calculate and plot the U.S. annual use of fuel wood, of coal, of petroleum, of natural gas, of hydropower, and of nuclear energy from 1850 to 2000.

7.4

Compare the ratio of air conditioning to space heating energy use in the commercial sector with the same ratio in the residential sector. Why are they so different? Similarly, compare the ratio of cooking to refrigeration energy use in commercial and residential sectors, and comment on the differences.

7.5

How does the energy used by the food and kindred products industrial classification (Table 7.3) compare with the caloric value of the end product (which approximately equals the U.S. food intake)?

7.6

What fraction of U.S. transportation is used for local transportation as compared to intercity transportation? What fraction is used for "moving people" as compared to "moving things"? (Note that uses of trucks for other than intercity freight will include use of light trucks, pickups, delivery vans, etc. Some fraction of this use is for moving people rather than things.)

7.7

From the information in Tables 7.2, 7.3, and 7.5, compare the energy content of petroleum with the energy required for refining it. If the energy used in petroleum refining is charged against the end uses of petroleum rather than against the industrial sector, how do the percentages used by the four end-use sectors change?

7.8

Attempt a determination of your total annual energy consumption by categories. Wherever possible, consult records (utility bills, etc.). Where such direct records are not available, use indirect information such as miles driven (from odometer) and estimated gas mileage of car. Finally, for areas where even less information is available, make educated guesses (and justify them). How do your energy uses compare with the norm for residential and passenger transportation uses? Make conjectures as to reasons for any significant differences.

7.9

Make a quantitative determination of the flows of energy through some well-defined entity such as a home, a dormitory, a university, an office building, a manufacturing plant, or a political division (city, town, rural county). As in problem 7.8, use information as direct as can be obtained. Compare your findings with national norms where appropriate and comment on any significant differences. (Because of the effort involved in obtaining the data, this problem constitutes a major term project.)

7.10

Make a comparative study of energy in a few different countries considering both sources and end uses. For each country, obtain the sorts of information that are given for the United States throughout Chapter 7. (Because of the effort required to obtain the data, this problem constitutes a major term project.)

7.11

Determine the retail cost in your community of various sources of energy. Note how the costs have changed over the past several years. For natural gas and electricity, note the rate structure. Compare the cost per unit of gas and electricity for a typical small consumer with that for a typical large consumer. (Throughout, express costs both as cost/gal, cost/ton, etc., and as cost/10^6 Btu.)

7.12

Compare the cost of heating a home using natural gas at $1.65/1000 cu ft *vs.* using electricity at $0.015/kWh. Consider utilization efficiencies in your calculation. Compare costs for one year for a typical home.

7.13

What will the U.S. annual energy use be in the year 2000 if the projected growth rate of 2.5 percent per year proves to be correct? if the projected rate of 4 percent proves correct? What will the total energy requirements from 1975 to 2000 for the United States be in each case?

7.14 *

Attempt a projection of U.S. per capita residential energy use assuming that there will be no significant new technological changes or life-style changes, but that energy use growth resulting from recent technological changes and life-style change trends will continue, with each individual component (e.g., air conditioning, frost-free refrigerators) growing at its present rate until saturation occurs (e.g., until all homes are air conditioned). With these assumptions, per capita residential energy use growth will eventually level out. This problem asks you to estimate when and at what level this will occur.

a) In making your projection, consider each component of residential energy use separately. Wherever you lack appropriate data, specify what data are needed and then make a plausible guess.

b) Plot total per capita residential energy use, as projected, from the present until it levels out. Compare the ultimate distribution among the various categories (i.e., space heating, cooking, etc.) with the present distribution.

c) The assumptions of the first paragraph were given in order to allow a projection to be made. Comment on their plausibility. Can you give an alternative set of assumptions that will also allow you to make projections?

Notes

1
H. C. Hottel and J. B. Howard, *New Energy Technology—Some Facts and Assessments*. Cambridge, Mass.: MIT Press, 1971, p. 219.
2
"A Review and Comparison of Selected United States Energy Forecasts," prepared for the Office of Science and Technology by Battelle Memorial Institute, December 1969.

Related Reading

Energy and Power, A Scientific American Book, September 1971 issue of *Scientific American*. San Francisco: W. H. Freeman, 1971. Contains several articles on energy use by different societies: hunting, agricultural, and industrial.

Hans H. Landsberg and Sam H. Schurr
Energy in the United States—Sources, Uses, and Policy Issues, A Resources for the Future Study. New York: Random House, 1968.

Sam H. Schurr and Bruce C. Netschert
Energy in the American Economy, 1850–1975, Resources for the Future, Inc. Baltimore: Johns Hopkins Press, 1960.

SRI
Patterns of Energy Consumption in the United States, a study for the Office of Science and Technology by Stanford Research Institute. Washington, D.C.: U.S. Government Printing Office, 1972.

NAE
Engineering for Resolution of the Energy-Environment Dilemma, Report of the Committee on Power Plant Siting, National Academy of Engineering, Washington, D.C., 1972. Contains an interesting section entitled, "Energy and Economic Growth."

Sam H. Schurr (ed.)
Energy, Economic Growth, and the Environment, Papers presented at a Forum conducted by Resources for the Future, Washington, D.C., April 20–21, 1971. Baltimore: Johns Hopkins University Press, 1972. A long appendix by Joel Darmstadter contains much useful data.

W. E. Morrison and C. L. Readling
An Energy Model for the United States, Featuring Energy Balances for the Years 1947–1965 and Forecasts to the Years 1980 and 2000. U.S. Department of the Interior, Bureau of Mines, No. 8384, 1968.

U.K. Energy Statistics, Her Majesty's Stationery Office, London, 1973.

8 Energy Conservation and Growth Reduction

Use it up;
Wear it out;
Make it do;
Do without.

Old New England Proverb

8.0 Introduction

The projections given in the preceding chapter for continued rapid energy growth imply some severe problems for the future—resource depletion, balance-of-payments deficits, aggravation of energy-associated environmental problems, fuel shortages, and so on. Indeed many of these problems are already upon us. It is clear that the growth in human energy use must stop sometime, probably sometime soon. In this chapter we will consider ways to use less energy, ways to slow energy growth. Approaches can be grouped into the categories of energy conservation and life-style changes.

The term *energy conservation* implies accomplishing essentially the same objective as heretofore (e.g., heating a house, producing a ton of steel) but using less energy to do so. This can be done by (a) increasing the use of some other resource (i.e., materials or labor), (b) settling for a slightly lowered objective, or (c) being smarter (i.e., inventing new techniques, using new technology, giving more thought to saving energy). As an example, the energy required for heating a home can be lowered by (a) installing more insulation (materials and labor), (b) setting the thermostat slightly lower (lowered objective), or (c) orienting the house and its windows so that more sunlight enters during the winter months (being smarter). The term *life-style change* implies altering the objective significantly. Living in a small apartment rather than a large single-family dwelling is a life-style change that lowers space-heating energy needs.

In the next two sections we will consider space heating and transportation in some detail. These two uses are singled out because they are by far the two largest end uses, accounting for close to half of current energy consumption. Furthermore, all individuals are directly involved in these two uses. (Whereas the industrial sector uses very large amounts of energy, there is no single use that stands out.) In Section 3 other uses will be considered, and in Section 4 some general observations and conclusions will be presented.

8.1 Space Heating

Space heating is used to maintain the interior temperature of a building (home, store, factory) at a level higher than the ambient outside temperature. Energy is required to replace the heat lost by the building. These losses occur via conduction of heat through the exterior wall surfaces, roof, and windows, and via exchange of air between interior and exterior through cracks around doors and windows (and through open doors and windows) and through cracks and pores in the walls and roof themselves. All these losses are directly proportional to the temperature difference between the interior and exterior of the building. An expression for the heat losses is given in Eq. 8.1:

$$\frac{dE}{dt} = \left(T_{in} - T_{out} \right) \cdot \left(\sum_j A_j U_j + cVn \right) \tag{8.1}$$

The sum is carried out over all exterior surfaces (walls, roof, windows). A_j is the area of each surface and U_j is the heat transfer coefficient of each surface; c is the specific heat of air (0.018 Btu/ft^3-°F); V is the volume of the house; n is the number of complete air changes per unit time. Heat losses can be reduced by reducing any of the terms in Eq. 8.1.

Values of U for different surfaces with different amounts of insulation are shown in Table 8.1. Insulation reduces the U values significantly. Note the considerably higher U value of a window as compared to a well-insulated wall or ceiling.

Infiltration of outside air can be reduced by weatherstripping around windows and doors, by use of storm windows, and by keeping windows and doors closed. However, some infiltration is desirable for purposes of ventilation. A reasonable value to assume for n is 0.5/hr, i.e., an air change every two hours. Values of both n and U_j increase if there is a strong wind.

The hourly heat losses for a "typical" home (30 ft × 60 ft single story, ranch style, with 260 sq ft of windows and unheated basement) with an inside-to-outside temperature differential of 70°F under assumptions of minimal insulation and heavy insulation are shown in Table 8.2. Note the importance of window and infiltration losses, particularly in the well-insulated case.

Table 8.1 U values for different surfaces with various amounts of insulation.

Component	Insulation	U (Btu/ft²-hr-°F)
Ceiling	None	0.61 (0.44)(0.14)[a]
	1⅞-in. glass fiber	0.12 (0.11)(0.072)
	2½-in. glass fiber	0.092 (0.087) (0.061)
	3½-in. glass fiber	0.068 (0.066)(0.050)
	6-in. glass fiber	0.042 (0.041)(0.034)
Wall	None	0.19
	1⅞-in. glass fiber	0.082
	2½-in. glass fiber	0.069
	3½-in. glass fiber	0.055
Floor	None	0.28
	1-in. glass fiber	0.138
	2-in. glass fiber	0.091
	Foil with air gap	0.093
Window	Plain	1.02 + 0.44[b]
	Double glazed,	
	¼-in. air gap	0.58 + 0.44
	Storm window	0.48 + 0.22

[a] Values in parentheses are for heat flow downward (during cooling). First value is for ceiling; second value includes effect of attic and roof.

[b] First value is conduction U; second value is infiltration equivalent U calculated on the basis of a 3-ft × 5-ft window. Effective U is the sum of the two values.

Source: John C. Moyers, "The Value of Thermal Insulation in Residential Construction: Economics and the Conservation of Energy," ORNL-NSF-EP-9. Oak Ridge National Laboratory, Oak Ridge, Tennessee, December 1971.

Table 8.2 Heat losses from a "typical" house with an inside-to-outside temperature differential of 70°F, assuming minimal insulation (3½-in. glass fiber in ceiling, none in walls or floor, plain windows) and heavy insulation (6″ glass fiber in ceiling, 3½″ glass fiber in walls, foil with air gap under floor, storm windows). In addition to infiltration around windows, a general infiltration of one air change per two hours is assumed for both cases.

Component	Minimal insulation		Heavy insulation		Percent reduction due to heavy insulation
	Heat loss (Btu/hr)	Percent of total	Heat loss (Btu/hr)	Percent of total	
Ceiling	7,674	10.6	4,937	13.5	36
Walls	15,694	21.6	4,543	12.4	71
Floor	12,600	17.3	4,185	11.5	67
Windows	26,572	36.6	12,740	34.9	52
General infiltration	10,080	13.9	10,080	27.7	—
Total	72,620	100.0	36,485	100.0	50

Source: John C. Moyers, "The Value of Thermal Insulation in Residential Construction: Economics and the Conservation of Energy," ORNL-NSF-EP-9. Oak Ridge National Laboratory, Oak Ridge, Tennessee, December 1971.

John Moyers has studied the monetary cost or savings and the energy savings that would result from application of varying amounts of insulation to a residence.[1] Comparison was made with the Federal Housing Administration minimum property standards, which are those standards a house must meet to qualify for an FHA-insured mortgage. Moyers finds that the economic optimum amount of insulation (that which results in lowest overall dollar costs to the homeowner) is substantially more than is required by the pre-1971 FHA standards, and slightly more than is required by the FHA standards as upgraded in 1971. He suggests a further upgrading of insulation requirements.

Heat losses per residence can be reduced by making the residences smaller (smaller A and V). Heat losses per household can be reduced by including more than one household in the same building (as in duplexes and apartment buildings), thereby reducing the exterior surface per unit of living space. Reducing the amount of window area cuts heat losses, but also increases lighting requirements and reduces heat input from sunlight. Window heat losses can be reduced somewhat by closing the drapes at night.

$T_{in} - T_{out}$ can be reduced by either lowering the inside temperature or moving to a warmer climate. The annual average for $T_{in} - T_{out}$ is usually stated in terms of *degree-days*—a sum over the days of the heating season of the day-to-day average of $(65 - T_{out})°F$. The degree-day, while based on 65°F, can be used for any desired interior temperature as long as the average length of the heating season is known. (It is assumed that with an outside temperature above 65°F, heating is not required.) The degree-days per year for Atlanta, New York, and Minneapolis are 2826, 5050, and 7853, respectively. Heating season average T_{out} values are 50°F, 44°F, and 35°F, implying heating seasons of 188, 240, and 261 days, respectively. Lowering the thermostat from 70°F to 65°F would save 25 percent, 19 percent, and 14 percent of the heating losses for residences in the three cities, respectively.

The heat losses given by Eq. 8.1 are replaced largely with energy from the home heating unit (i.e., the furnace), but also with energy from other sources. For example, human occupants generate 100 watts per individual, and the use of electrical power for other than space heating is on the order of 1000 watts per residence. Thus a home with four people in it yields 1400 watts or 4800 Btu/hr from these sources, which is about 25 percent of the average heating requirements of a well-insulated home. Solar radiation can also be important and can be increased by architectural design, e.g., window location, and the orientation of the house.

The efficiency of a gas furnace in good repair is about 85 percent; an oil furnace in good repair is 65 percent efficient. With poor furnace maintenance, these efficiencies fall.

There has been considerable debate about whether or not the shift toward electric heating is a "good thing." It is appropriate at this point to list the relative merits of electric resistance heating and direct fuel use in furnaces for residential space heating.

1. With 75-percent home-furnace efficiency, 100 percent electric resistance heating efficiency, and 35 percent generating efficiency, fuel requirements for

direct use of fuel are just under half those of electric resistance heating. If other factors were equal (which they aren't, see below), the environmental ill effects from extracting, transporting, and using the fuel would be only half as large with direct use as with electric generation.

2. With direct use of fuel, air pollutants are dispersed over a wide area, being released in small amounts from many homes rather than in large amounts from a single point (the power plant). This dispersal is good, but the fact that the air pollutants are released in a residential area rather than elsewhere is bad.

3. With direct use of fuel, there is no release of waste heat to a body of water, i.e., no thermal pollution in the conventional sense.

4. Taking advantage of economies of scale, a central power plant can have more effective air pollution abatement techniques (tall stacks, electrostatic precipitators) than can individual home furnaces.

5. Good maintenance is more readily achieved in a central power plant than in many individual furnaces. (Resistance heaters require negligible maintenance.) Without good maintenance, efficiency falls and air pollution increases.

6. Central power plants can use some energy sources not available to home furnaces, i.e., coal, uranium, and water power.

The "winner" in the comparison is not obvious and will vary from situation to situation. An ample supply of hydroelectric power argues for electric heating; if natural gas is the fuel, it should be used directly rather than via electricity; a serious aquatic thermal pollution problem argues for direct use of fuel; if coal must be used, electric heating is called for; etc. Perhaps the "winner" should be an entry not really in the race yet—the electrically driven heat pump. This approach has most of the advantages listed for electric resistance heating (all except better maintenance) but has fuel economy comparable to home furnaces (see Section 5.6).

8.2   Transportation

Transportation refers to the movement both of people and of goods and materials. The standard measures of energy efficiency for the two cases are passengers times distance per unit energy (passenger-miles per gallon of fuel) and freight units times distance per unit energy (ton-miles per gallon of fuel). Though weight is the freight characteristic of primary importance, volume is also relevant—it takes more energy to move a ton of very low-density material occupying a small volume.

Energy efficiencies for various modes of intercity freight transport are listed in Table 8.3. Because the high-efficiency modes (waterways and railroads) are traditionally used for high-density materials (e.g., coal, ore) and the low-efficiency modes (trucks, airways) carry a higher percentage of low-density, bulky items, the table probably slightly overstates the difference in efficiencies. Nevertheless, the differences are substantial.

Energy efficiencies for various modes of intercity and urban passenger traffic are listed in Table 8.4. Since the energy requirements of most vehicles depend

Table 8.3 Energy efficiency for intercity freight transport during 1970.

Type of transport	Ton-miles per gallon	Share of ton-miles in 1970 (percent)
Pipelines	300	22.4
Waterways	250	15.9
Railroads	200	40.1
Trucks	58	21.4
Airways	3.7	0.18

Source: Eric Hirst, "Energy Consumption for Transportation in The United States," ORNL-NSF-EP-15. Oak Ridge National Laboratory, Oak Ridge, Tennessee, March 1972.

Table 8.4 Energy efficiency for intercity and urban passenger traffic during 1970.

Type of transport	Passenger-miles per gallon	Share of passenger-miles in 1970 (percent)
Intercity		
Buses	125	2.1
Railroads	80	0.9
Automobiles	32	87.0
Airplanes	14	9.7
Urban		
Buses	110	4.6
Automobiles	27	95.4

Source: Eric Hirst, "Energy Consumption for Transportation in the United States," ORNL-NSF-EP-15. Oak Ridge National Laboratory, Oak Ridge, Tennessee, March 1972.

very little on how many passengers they are carrying, the efficiencies are directly proportional to the load factor (fraction of full capacity). Load factors are frequently low, averaging perhaps 50 percent. It is evident from the table that most passenger transportation uses low energy-efficiency modes rather than high energy-efficiency modes, and that the differences among the modes are substantial.

Of course, energy efficiency is not the only consideration in selecting a mode of transportation. Other considerations include speed, comfort, dependability, safety, convenience, and cost. Certain modes of transportation do not apply in some situations—water transport between Chicago and Denver, railroads between New York and London, pipelines as a transportation mode for fresh strawberries, etc.

The automobile plays by far the dominant role in passenger transportation. The distribution of auto trips for different purposes is given in Table 8.5.

It is of interest to determine the reasons for the level of energy consumption of various transportation devices. We shall focus on the automobile. A sizable inefficiency occurs in the conversion of the chemical energy of gasoline to

Table 8.5 Distribution of passenger car trips, travel, and occupancy by major purpose of travel, 1969–1970.

Purpose of travel	Percentage distribution		Average trip length (miles one-way)	Average number of occupants per car
	Trips	Travel		
Earning a living				
To and from work	32.3%	34.1%	9.4	1.4
Business related to work	4.4	8.0	16.0	1.6
Subtotal	36.7	42.1	10.2	1.4
Family business				
Medical and dental	1.8	1.6	8.3	2.1
Shopping	15.4	7.6	4.4	2.0
Other	14.2	10.4	6.5	1.9
Subtotal	31.4	19.6	5.5	2.0
Educational, civic, or religious	9.4	5.0	4.7	2.5
Social and recreational				
Vacations	0.1	2.5	165.1	3.3
Visit friends or relatives	9.0	12.2	12.0	2.3
Pleasure rides	1.4	3.1	19.6	2.7
Other	12.0	15.5	11.4	2.6
Subtotal	22.5	33.3	13.1	2.5
All purposes	100.0%	100.0%	8.9	1.9

Source: Preliminary results from the *Nationwide Personal Transportation Survey, 1969–1970,* Department of Transportation, Federal Highway Administration, Office of Planning. As reported in "The Potential for Energy Conservation—A Staff Study," Office of Emergency Preparedness, October 1972.

mechanical energy in the engine. When traveling at a constant speed, mechanical energy is dissipated because of air drag, tire friction, and internal friction in the engine and other moving parts. Energy is required for acceleration and is dissipated during braking. Energy is required for grade climbing, though it is recovered during the descent. Finally, while an auto idles in traffic, fuel is consumed. We will consider each item in turn.

An internal combustion engine is capable of an efficiency around 25 percent. This efficiency is obtained when the engine is running under optimum conditions of load, rate of fuel supply, and engine rpm, i.e., at a given power level and speed. For conditions other than the optimum, naturally the efficiency will be less. The automobile engine is subject to a wide variety of demands; it must be capable of high power for rapid acceleration and low power for constant-speed driving. In spite of the transmission, the engine speed varies over a substantial range. Therefore it is to be expected that the average engine efficiency will be significantly below the optimum efficiency.

The engine efficiency of an automobile having a manual transmission driven at constant speed on level terrain can be termined as follows. First determine the gasoline mileage under the above-mentioned conditions, and convert the result to energy units. Then determine the rate at which the automobile slows down when the clutch is depressed. From this and the mass of the car, calculate the energy dissipated per mile "from the clutch on." The ratio of energy dissipated per mile to fuel energy per mile is the engine efficiency. (Friction losses in the engine are treated as a source of engine inefficiency in this method.) In this way, the author's Volkswagen Beetle was determined to have an engine efficiency of 16 percent at 55 mph. The car weighed 790 kg, obtained 35 mi/gal at 55 mph, and slowed from 60 mph to 50 mph in 9 seconds after the clutch was depressed. These numbers yield a fuel use of 3.8×10^6 joules/mi and an energy dissipation from the clutch on of 6.2×10^5 joules/mi.

We can crudely estimate the energy dissipated in air turbulence (the air drag) by assuming that the volume of air swept out by the vehicle (vehicle cross-sectional area A times distance traveled d) is given a speed equal to that of the car. Then the energy dissipated is $\frac{1}{2}\, \rho A d v^2$, where ρ is the density of air and v is the speed of the vehicle. A more refined treatment shows that the above expression must be multiplied by a coefficient of proportionality called the drag coefficient c_D, which depends on the shape of the vehicle and on the Reynolds number $R = v\sqrt{A}/\nu$, where ν is the kinematic viscosity of air, 0.15 cm^2/sec. Thus

$$\frac{E}{d} = \frac{1}{2}\, c_D\, \rho A v^2. \tag{8.2}$$

For Reynolds numbers above 300,000 (speeds above 10 mph for a vehicle with dimensions like an automobile, traveling in air), flow is turbulent and c_D depends very little on R. For a sphere, $c_D \approx 1/5$. A reasonable value to use for an automobile is $c_D = \frac{1}{2}$. Note the dependence of air turbulence losses on vehicle size and speed; cross sectional area and speed squared. Evaluating Eq. (8.2) for a Volkswagen beetle at 55 mph ($A = 1.5$ m^2), we find a dissipation of 4.5×10^5 joules/mi, which is about 70 percent of the losses observed for the author's car.

Mechanical energy is dissipated internally in an automobile via friction in the transmission, differential, and wheel bearings. (For convenience in measurement, we have defined those losses that occur upstream of the clutch (pistons, main bearings) as being part of the engine inefficiency.) There is no easy way to measure or calculate these losses. They are roughly proportional to the weight of the automobile and depend little on automobile speed.

Mechanical energy is converted to heat at the interface between road surface and tire and inside the tires due to flexing. Again, there is no easy way to measure or calculate these losses. They are proportional to the weight of the automobile and depend little on its speed. Tire losses and internal friction must account for those power losses in excess of air turbulence and are therefore approximately 30 percent of the losses at 55 mph for the author's Volkswagen.

The energy required to accelerate an automobile of mass m to a speed v from rest is $\frac{1}{2}\, mv^2$. Assuming an engine efficiency of 16 percent, 0.013 gallon of gas is required to accelerate a 790-kg car from rest to 60 mph. In urban driving with many stops and starts, the acceleration energy requirement can add up. Ten accelerations per mile from rest to 30 mph would require 0.033 gallons per mile for acceleration alone. (This is rather extreme stop-and-go driving.)

The energy required to climb a hill of height h is mgh. For a 100-foot hill, this is the same energy as is required to accelerate from rest to 55 mph on level road. Assuming it is not necessary to brake during the descent, the energy used in climbing a hill is "recovered" in descending it.

Fuel is consumed in an automobile engine as long as it is running, even though no power is being delivered. Specifically, when the car is stopped in traffic and the engine idles, enough energy must be supplied to keep the engine "turning over" against engine friction. This power requirement is roughly proportional to the mass of the engine, which in turn is roughly proportional to the power of the engine, which in turn is roughly proportional to the mass of the automobile.

From the preceding discussion, we can understand most of the differences in energy efficiencies listed in Tables 8.3 and 8.4. For example, with its great length per unit cross-sectional area, a train experiences much less air resistance per unit carrying capacity than a truck or automobile. (For a train, c_D will be larger than for a shorter vehicle, but not that much larger.) With steel wheels on steel rails, "tire" losses are also much less. Internal friction is the principal use of energy at constant speed. Because "traffic" problems are much fewer, accelerations and idling are less frequent. Finally, because rapid acceleration is not required, the engine need not be so overpowered at steady running speeds, and so can operate nearer to optimum efficiency more of the time. In addition to these features, there are "economies of scale." Larger vehicles such as buses have more carrying capacity per unit vehicle weight than automobiles, with correspondingly higher energy efficiencies.

To the energy use discussed above we must add nontransportation uses for transportation vehicles. For example, air conditioning in automobiles adds significantly to their gasoline consumption.

Some industrial and commercial energy uses are very closely related to transportation. Energy used in petroleum refining, automobile manufacture, and road construction is considerable. The energy used in petroleum refining amounts to 12 percent of the energy content of the end product. The total energy requirements for the manufacture of an automobile (including producing the raw materials) is equivalent to 1000 gallons of gasoline, i.e., about one year's consumption by that automobile. By considering these and other energy uses related to the automobile (lubricating oil and tire manufacture; parking and garaging; insurance; repairs, maintenance, and parts; retail selling of automobiles and related items), Hirst[2] estimated that cumulatively they amount to 60 percent of the direct fuel use, and that the automobile is therefore responsible for 21 percent of all energy use in the United States.

Changes in transportation energy use arise because of changes in the energy efficiency of the various modes, shifts in percentage use among the various modes, changes in per capita passenger miles or freight miles, and changes in population. The relative roles of these factors in the increase in transportation energy use between 1960 and 1970 are shown in Fig. 8.1. The poorer energy efficiency of passenger transportation was due both to a small deterioration in energy efficiency of automobiles and to a shift to less efficient transportation modes (from bus and

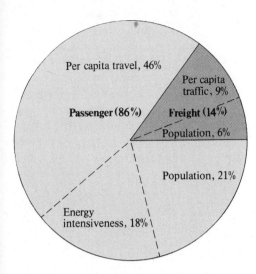

Fig. 8.1 Factors accounting for the increase in transportation energy use between 1960 and 1970. (From *Energy Conservation,* Hearings before the Committee on Interior and Insular Affairs, U.S. Senate, 93rd Congress, March 1973, p. 163.)

rail to automobile and airplane). In freight transportation, improved energy efficiency of railroads barely compensated for a shift to less efficient modes (trucks and airplanes). Use of airplanes both for passengers and for freight is growing very rapidly; if this trend continues, it will be the dominant cause of increased transportation energy use in the next few decades.

Because U.S. transportation is so grossly wasteful of energy, it is easy to devise schemes that would greatly reduce energy use. However, implementation of these schemes is usually not easy. Fundamental to the problem is the pattern of living in the United States, i.e., suburbia. With single-family residences spread diffusely over a large area, and each residence generally a great distance from one's place of employment and invariably not within easy walking distance of public areas or stores, transportation energy requirements are bound to be high. Not only is an automobile necessary for commuting to work and for shopping, it is necessary for almost every activity including many children's programs (Scout meetings, music lessons, etc.). The suburban housewife often has to drive more miles per day than her husband. The large increase in per-capita travel shown in Fig. 8.1 is due substantially to increased distances in commuting, shopping, and other tasks. Therefore, this portion of the increase represents a decrease rather than an increase in standard of living or quality of life.

The pat answer for urban passenger movement is mass transit—buses, subways, rail lines, etc. Table 8.4 shows a factor of 4 or more improvement in energy efficiency for these modes over automobiles. Indeed, mass transit systems work very well in European cities. However, the land-use patterns described in the preceding paragraph make mass transit a difficult proposition in most U.S. metropolitan areas. Not only must passengers be collected from a large, low-density region (suburban residential areas), but the points to which they must travel are becoming increasingly more diffuse as industrial and commercial establishments leave center-city areas for the suburbs. Motivated by the desire to relieve urban air pollution and traffic congestion, many U.S. cities are attempting to revitalize their mass-transit systems and are using such devices as "park-and-ride" and "dial-a-bus" to overcome the problems of suburbia. Yet it appears unlikely that mass transit will be able to satisfy the bulk of the transportation needs of most U.S. urban areas without substantial changes in land-use patterns.

The pat answer to intercity passenger movement is railroads or intercity buses. Again, this approach (railroads) works very well in Europe. However, in the United States, the railroads have lost their short-haul passengers to the automobile and their long-haul passengers to the airlines. Various government policies of taxation, regulation, and subsidy have accelerated the trend. A major obstacle to use of railroads or buses for short-haul intercity travel is that an automobile is invariably required both to travel to the terminal to catch the bus or train and also to conduct one's business in the destination city. Unless the urban mass-transit problem can be solved, intercity public transportation will remain at a considerable disadvantage relative to the automobile for short-haul trips ($\leqslant 100$ miles). For long-distance trips, the advantage of the airplane's shorter travel time is often overriding.

Starting from scratch, one can design urban areas whose transportation requirements are considerably less than those of current U.S. cities (by factors of 4 or more) and whose quality of life is higher. The essential component is clustering. A typical design might consist of a high-density center city (with commercial, industrial, and residential facilities) surrounded by satellite towns. These towns are high density compared to present-day suburbia (duplexes, apartment buildings, and some single-family dwellings) and include stores and similar commercial facilities. The space "freed up" by the clustering remains as open space —farmland, parkland, etc. Thus the desire to "get back to nature" is satisfied better than in suburbia. A large fraction of an individual's activity would go on within the satellite town in which he lived or in the center city. Much transport within a satellite town would be by foot or bicycle, typical distances being under one mile, while transport from town to center city and within the center city would be by mass transit (quite feasible at the population densities in the towns and the city).

While the above description of an urban area sounds better to most people than what we now have, it's not at all clear that we can "get there from here." Suburban sprawl is difficult to reverse.

Aside from altering land-use patterns or struggling for mass transit in spite of them, what can be done? Use of smaller, lighter, lower-powered automobiles can save as much as a factor of 2 in energy consumption; in most cases, automobiles are larger and more powerful than necessary. Carpooling can increase the load factor in automobiles; the present average number of occupants per car in urban driving is only 1.2. With some advanced planning and thought, the number of trips devoted to household errands, etc., can be reduced. The incentive to do any of the above is most readily created by an increase in gasoline price.

Nothing has been said here about energy reductions that might result from new technology—more efficient engines, etc. I am doubtful that there will be any large gains here; the answers to America's excessive transportation energy use are social, not technical.

Unlike the passenger transportation situation, there is no fundamental obstacle to increasing the amount of freight that is carried by energy-efficient modes, that is, to switching freight from trucks to rails. In particular, the "piggy-back" approach, carrying truck trailers on flatcars, can simplify the pickup and delivery problems. With routing handled by computers, there is no reason why freight shipped by rail cannot arrive as quickly as freight shipped by truck.

8.3 Other Areas

Industry

It is reasonable to assume that industry is very cost conscious and that all industrial processes are near optimum efficiency *at current prices and tax structures.*[3] This means that industry could not produce the same product with less energy and the same amount of other ingredients. If the price of energy were significantly raised, industry would reoptimize its processes, using less energy and more of something else (labor, materials, capital). A recent study by the Office of Emergency Preparedness suggests that industry could save 15 to 20 percent of its energy needs (given appropriate financial incentive) by keeping plants modernized (capital) and by keeping equipment well maintained (labor).[4]

In many cases, recycling of materials can reduce industrial energy requirements as well as conserve nonenergy resources and ease solid-waste disposal problems. For example, pulp (for paper products) made from recycled waste paper uses only one-fourth as much steam energy and one-tenth as much electrical energy as pulp made from wood.[5] Aluminum made from scrap requires only one-ninth as much energy as aluminum made from ore.[6] These figures do not include energy requirements for mining ore and harvesting pulpwood, or for separating scrap aluminum and waste paper from other solid waste. Transportation energy also is not included. It appears likely that more complete studies including these factors would still show substantial energy savings in using recycled materials. Through tax policies, transportation rates, procurement policies, and other devices, government can favor use of either virgin or recycled materials. The present bias favors virgin materials and is therefore a poor policy from the energy-conservation and resource-conservation points of view.

In the same vein, some products currently being discarded after a single use could be reused several times with resultant energy savings. Most examples fall in the general category of containers (boxes, bottles, bags); the beverage container (for soft drinks, milk, and beer) heads the list. Hannon[7] has compared the energy requirements for reusable as opposed to throwaway beverage containers, considering the entire system including acquisition of raw materials, transportation, and disposal of discarded containers. He finds that three times more energy is required to deliver a unit of beverage to the consumer by using throwaway bottles or cans than by using returnable bottles.

Similarly, increasing the usable lifetime of durable goods can reduce industrial energy use, because there is a longer period of time over which to "depreciate" the energy used in manufacture of the goods. In other words, it will be longer before a replacement is required.

Air Conditioning

Much of the discussion of space heating (Section 8.1) is relevant to air conditioning in that improved insulation keeps heat out in the summer as well as keeps it in during the winter. Setting a thermostat high in the summer helps in the same way as setting it low in the winter. The input of solar energy through the windows, which is desirable for space heating, is undesirable for air conditioning because it represents a major source of heat influx. Clever architectural design, such as an overhanging roof which blocks the sun from a window in the summer but lets it in during the winter (because the sun is then lower in the sky), or deciduous trees located so they block the sunlight in the summer when they are in leaf but let it through in the winter when the leaves are off, can be used to minimize energy requirements in both cases.

The efficiency of air conditioners varies considerably from model to model; 1972-model room air conditioners ranged from 5 to 12 Btu of heat removed per watt-hour of electrical energy supplied.[8] Twelve Btu/watt-hr is an efficiency of 3.5, to be compared with a Carnot efficiency limit $T_1/(T_2 - T_1)$. With T_1 equal to 300°K (27°C, or 81°F) and $T_2 - T_1$ equal to 10°K (18°F), the Carnot limit is 30. Therefore, even the units with the highest efficiency have plenty of room for improvement.

Home Appliances

Most home appliances are not designed with much attention given to efficiency, and they can be easily improved in that respect. Refrigerators and hot-water heaters could benefit from better insulation, for example.

Generation of Electricity

As discussed in Chapter 5, fossil-fuel-fired, steam-electric generating plants have an overall efficiency of only 40 percent, whereas light-water nuclear reactors have an even lower efficiency of 33 percent. Any scheme that would raise the efficiency of electricity generation would result in energy saving. Any scheme that would utilize the waste heat from a power plant would similarly reduce energy use. Both approaches would also reduce thermal pollution.

In the former category, systems combining a gas turbine at very high initial temperatures with a steam turbine at contemporary initial temperatures promise efficiencies in excess of 50 percent. And magnetohydrodynamic schemes promise similarly improved efficiencies.[9]

The most obvious use of the low-temperature waste heat from a power plant is for space heating. This requires locating the power plant in a densely populated area. The turbine must be operated at a somewhat higher outlet temperature than normal, with resulting reduction in efficiency of electricity generation, in order to have the waste heat at a sufficiently high temperature to be useful. A distribution system of pipes is required to carry the hot water or steam from the power plant to commercial or residential users. One obvious defect in such an approach is the seasonal nature of the space-heating requirement. Industrial uses, less seasonal in nature, generally require high-temperature waste heat.

8.4 Observations and Conclusions

The potential energy savings that various conservation measures could accomplish have been investigated by the Office of Emergency Preparedness, assisted by staff members of many other agencies. Their study[4] suggests that a strenuous program of energy conservation could reduce energy growth between 1970 and 1990 by a factor of two, reducing energy growth from a doubling to a factor of 1.5 increase. Improvements in space heating and cooling (principally more insulation), reduced reliance on automobiles (mass transit, new transportation systems), improved industrial processes, and recycling were the major sources of savings.

It must be understood that essentially all energy conservation measures involve tradeoffs. One is trading energy savings for use of materials or labor, or for loss of convenience, performance, or comfort. The savings in energy per "price" paid are quite variable, but often very large. For example, saving a factor of two in space-heating energy by lowering the thermostat to 45°F, halfway from the normal heating temperature of 70°F, to an outdoor ambient of 20°F, involves an unreasonable "price" in discomfort. By contrast, saving a factor of two in commuting transportation energy by carpooling with another individual involves a small "price" in loss of convenience. In energy-conservation matters, as with anything else involving tradeoffs, one must perform a cost-benefit analysis, balancing costs and benefits of all sorts, economic, environmental, and other.

Some energy-conservation measures would result in a monetary savings with no loss in performance. The major example is residential space heating. The Moyers study[1] demonstrates economic advantage to using more thermal insulation than is customary. Other examples are refrigeration and air conditioning; the increased purchase price of more efficient units is more than compensated for by reduced operating costs. Why are those measures that save both energy and dollars not carried out? The answer lies in how the costs are distributed; the energy-conserving measures result in higher first costs but lower operating costs. Whether buying a house or a refrigerator, the consumer is painfully aware of the first costs (the purchase price) but often is not supplied with sufficient information on operating costs to properly include them in his or her decision. Consumer protection laws requiring disclosure of operating costs and energy requirements would correct this, and thus can be viewed as measures for energy conservation as well as consumer protection. In the case of a commercial or industrial decision, tax and other policies often favor a quick writeoff of a capital expense, thus giving too much weight to first costs and discouraging conservation measures.

The increased use of energy in the United States is due to increases in population and to increases in per-capita energy use, both being of comparable importance. The increasing per-capita energy use is caused by life-style changes in the "wrong" direction. A life-style *nonchange* (e.g., a stable number of miles traveled per year, stable materials consumption per year) would decrease the rate of energy growth. (One might argue that "rising expectations" and increasing consumption have become a life-style, and that maintaining a constant per-capita energy use is a life-style change.)

Life-style changes that reduce per-capita energy use do not necessarily lower the quality of life. Since quality of life is such a subjective, ill-defined concept, this statement is easy to make but also easy to dispute. Consider, for example, a life-style based on the land-use pattern described in Section 8.2—high-density center city surrounded by open space studded with high-density satellite towns; no suburbs—and compare this with present U.S. land-use patterns—suburbia and urban sprawl. The per-capita transportation-energy requirements of the high-density setup are considerably lower than those of the sprawl arrangement. Furthermore, the space-heating energy requirements of the high-density setup can be lower than those of the sprawl arrangement even with the same amount of floor space per capita. This is possible because duplexes and small apartment buildings have less exterior surface per unit floor space than single family dwellings do. Which land-use pattern fosters a higher quality of life? My subjective belief is that the high-density arrangement does. The harried suburban housewife who logs perhaps 10,000 miles per year doing all her household errands (shopping, driving children to lessons, doctor's appointments, Scout meetings, etc., and driving whenever she wants to do anything) likely will agree; the individual who has spent a year in one of the well-organized, smaller European cities (e.g., Geneva, Switzerland) likely will also agree.

In concluding this chapter, one must realize that while energy conservation measures can slow the rate of energy growth for a short period of time, they cannot arrest energy growth over the long haul. Indeed, present energy growth is not due to "anticonservation" measures. If energy growth is to stop and energy use is to stabilize, then it is clear that both population growth and per-capita energy growth must also stop. For a while, conservation measures might reduce per-capita energy use sufficiently to compensate for increases due to greater consumption of goods, increased travel, etc. Eventually, however, a constant per-capita energy use will require no increase in material goods per capita, no increase in per-capita travel. GNP per capita must either remain constant or grow only by a shift from energy-intensive products to those requiring less energy (services and handcrafts, for example). (The concept of a stationary economy is foreign to the thinking of most economists and repugnant to all businessmen. We will consider the concept further in Chapter 9.)

Questions and Problems

8.1
Derive Equation 8.1.

8.2
In the third paragraph of Section 8.1, it is stated that both n and U_j increase in a strong wind. Explain why this is so.

8.3
Verify the heat-loss figures of Table 8.2 using the insulation figures of Table 8.1. For floor losses, assume the unheated basement is at a temperature 25°F below the inside temperature. (Due to some minor additional assumptions made by Moyers, you should not expect to obtain exact agreement in all cases.)

8.4
Calculate and compare the heating requirements for residences of different geometries: (1) a single-story ranch, 30 ft × 60 ft, (2) a two-story home 30 ft × 30 ft, (3) a two-story duplex home, 30 ft × 60 ft. (In all three cases, the amount of living space per family is the same, namely, 1800 sq ft of floor space.)

8.5
Calculate the annual heating requirement (Btu/yr) for the well-insulated house described in Table 8.2 for an inside temperature of 70°F and climatic conditions as given for Atlanta, New York, and Minneapolis. Justify the numbers given in Section 8.1 for percentage fuel savings accomplished by maintaining the inside temperature at 65°F rather than at 70°F.

8.6 *
a) Using some residence whose construction and heating performance you are familiar with, calculate the heat loss from the construction information along the lines of Table 8.2. For some heating season (one-year period, summer to summer), compare the calculated heat requirements with the actual ones obtained from utility bills. (Degree-day information can be obtained from the Weather Bureau.) In your comparison, remember to include an estimate of the furnace efficiency.

b) For the same residence, measure the rate at which the inside temperature falls when the furnace is turned off. Derive a formula for this rate of fall in terms of the inside-to-outside temperature difference, the constant describing the heat loss per degree-hour for the house, and the heat capacity of the house. Estimate the heat capacity from construction information. Compare the calculated and measured rates of temperature fall.

8.7 *
One strategy suggested for reducing the heating energy requirement is to turn the thermostat down (e.g., from 70°F to 55°F) in the evenings (9:00 P.M.) and to turn it up in the morning (6:00 A.M.). Calculate the percentage energy savings that this strategy accomplishes. (For this problem, you will need to know the rate at which the house cools down when the furnace is turned off—see Problem 8.6. Assume that the

hourly temperature drop dT_i/dt is proportional to the difference between the inside temperature T_i and the outside temperature T_o, namely $dT_i/dt = -\alpha(T_i - T_o)$. For the author's house, $\alpha = 2.5\%$/hr.)

8.8 *

An automobile uses energy (1) to maintain speed against air drag, (2) to maintain speed against tire friction and friction internal to the automobile, (3) to accelerate, and (4) to climb grades.

a) Write down expressions for the *power* required for each of the four items in terms of the automobile's mass M and cross-sectional area A, the speed v, the acceleration a, and the angle of grade relative to horizontal θ. (You will need only one arbitrary constant. Determine it from the data given for the author's car in Section 8.2.) Write down expressions for the *energy* required to travel one mile at the speed v, to accelerate from rest to the speed v, and to climb a hill of height h.

b) For an automobile of some chosen M and A, plot as a function of speed (1) the power required to maintain a constant speed, (2) the power required to accelerate at $1/10$ g (the acceleration of gravity), and (3) the power required to climb a 4-percent grade. Use the standard units for automotive matters, namely, horsepower and mph.

c) For the automobile used in part (b), calculate and plot gasoline mileage at constant speed as a function of speed, assuming the engine efficiency is 20 percent, independent of engine rpm and power output. You will find that the calculated gas mileage at low speeds is fantastic, unrealistically good. This is because the assumption about engine efficiency is bad. When very small amounts of power are being delivered from a large engine, engine efficiency falls. An improved assumption is that engine efficiency rises linearly from 0 to 20 percent as power output rises from 0 to $\frac{1}{4} P_{max}$ (maximum power), and then remains near 20 percent as output power increases from $\frac{1}{4} P_{max}$ to P_{max}. Calculate gasoline mileage using this assumption. (To do this, determine P_{max} by requiring that the car have "reasonable performance," namely, maximum speed $\geqslant$ 80 mph, acceleration of 0.1 g at 65 mph, ability to climb a 4-percent grade at 65 mph.

d) As is seen in part (c) above, there is a conflict between good gasoline mileage at intermediate-to-low speeds on the one hand, and good "performance" on the other hand. (This conflict is partially resolved by designing an engine and transmission so that the efficiency is near maximum for as broad a power range as possible. The factor of 4 range assumed in part (c) is doing pretty well.) In present-day automobiles, the conflict is resolved largely in favor of performance. Assume instead that you resolve the conflict in favor of good gasoline mileage by designing a car with M and A as used above, but with a small, low-powered engine such that $\frac{1}{4} P_{max}$ is the power required to travel at a constant 30 mph. For engine efficiency, make the second assumption in part (c) above, i.e., efficiency rising from 0 at zero power to 20 percent at $\frac{1}{4} P_{max}$. What is P_{max}? (Compare with P_{max} in part (c).) Plot gasoline mileage vs. speed. Evaluate various measures of performance such as maximum speed on level terrain, time to accelerate from rest to 30 mph, maximum speed up a 4-percent slope, and maximum acceleration at 40 mph.

8.9

What parameter of an automobile largely determines its "city-traffic" gas mileage? What parameter largely determines "high-speed-turnpike" gas mileage? Justify your answers.

8.10

Obtain the automobile mass M and cross-sectional area A for several different models of automobiles. Predict the gasoline mileage of each under various driving conditions (e.g., heavy city traffic, suburban; high-speed turnpike).

8.11 *
For a test automobile available to you (e.g., your own), study some energy-related characteristics.

a) Determine energy dissipated per mile "from the clutch on" as a function of speed, by measuring the rate of deceleration *vs.* speed with the clutch depressed. Is this dissipative force adequately described by the formula $a + bv^2$? Determine the drag coefficient C_D from the v^2 term. Is it consistent with ½? Compare your value of a with those obtained by others making the measurement. Is a roughly proportional to M? (To give you one more data point, for the author's car,

$$\frac{1}{\Delta t} = 0.0365 + 0.0075 \left(\frac{v_{mph}}{20}\right)^2,$$

where Δt is the time in seconds required to slow from $v + 5$ mph to $v - 5$ mph.)

b) Determine maximum power P_{max} by measuring acceleration up a grade of known slope or maximum speed up a grade of known slope.

c) Determine gasoline mileage for constant-speed driving at several different speeds. Using this information in conjunction with that in part (a), determine engine efficiency and power output P at these speeds. Plot engine efficiency *vs.* P/P_{max}.

   Warnings: As with any experiment, you will want to repeat your measurements several times to be sure the results are reproducible. Be sure the car is completely warmed up before you begin measuring. Give safety due regard—have one person drive and another record data.

d) Instead of using an automobile, carry out parts (a) through (c) using some other vehicle. A bicycle is an interesting possibility. In this case, the "engine" is the rider. Part (c) is very difficult to do, but in principal it can be done by measuring exhaled CO_2 and correcting for basal metabolism.

8.12
Discuss the various desirable features of a system for moving people from place to place. Indicate which of these conflict with the desire to minimize energy use, and in each case show why.

8.13 *
Outline a transportation system (which can be a mixture of public and private transit, automobiles, buses, bicycles, pedestrians, etc.) for a metropolitan area with which you are familiar. Attempt to reduce energy requirements while maintaining other desirable features. Compare your system with the system presently in use. (This question constitutes a term project.)

8.14
The average electricity generating efficiency (exclusive of hydropower) is now about 33 percent. Assume this efficiency rises to 40 percent over the next ten years due to replacement of older, less-efficient plants with modern plants and due to introduction of new technologies mentioned in Section 8.3. (The trend toward nuclear plants, with their 33-percent efficiency, will tend to slow the rise in efficiency.) Further assume that use of electricity continues to grow at a rate of 7 percent per year. At what rate will the fuel requirements for electric energy generation grow? At what rate will the heat releases at generating plants grow?

8.15

Outline an overall strategy for energy conservation, including schemes for implementation. Discuss how the strategy would probably work in practice, mention objections that would be raised, and consider difficulties that would be encountered.

Notes

1

John C. Moyers, "The Value of Thermal Insulation in Residential Construction: Economics and the Conservation of Energy," ORNL-NSF-EP-9. Environmental Program, Oak Ridge National Laboratory, Oak Ridge, Tennessee, December 1971.

2

Eric Hirst, "Direct and Indirect Energy Requirements for Automobiles," ORNL-NSF-EP-64. Environmental Program, Oak Ridge National Laboratory, Oak Ridge, Tennessee, February 1974.

3

This assumption has been challenged. It appears that industry was somewhat off the economic optimum, in the direction of using too much energy, even before the energy price increases of the mid '70's. See Charles A. Berg, "Conservation in Industry," *Science* **184,** No. 4134, p. 264.

4

OEP, *The Potential for Energy Conservation,* Office of Emergency Preparedness. U.S. Government Printing Office, Washington, D.C., 1972.

5

SRI, *Patterns of Energy Consumption in the United States,* A Study for the Office of Science and Technology by Stanford Research Institute. U.S. Government Printing Office, Washington, D.C., 1972, p. 125.

6

Ibid., p. 115.

7

Bruce Hannon, "System Energy and Recycling: A Study of the Beverage Industry." CAC Document No. 23, Center for Advanced Computation, University of Illinois, Urbana, Illinois, March 1973.

8

Energy Conservation, Hearings before the Committee on Interior and Insular Affairs, U.S. Senate, 93rd Congress, Part I. U.S. Government Printing Office, Washington, D.C., March 1973, p. 171.

9

H. C. Hottel and J. B. Howard, *New Energy Technology—Some Facts and Assessments.* Cambridge, Mass.: MIT Press, 1972, pp. 281, 283.

Related Reading

General treatments of energy conservation, in addition to references 4 and 8 include the following.

David B. Large (ed.)

Hidden Waste—Potentials for Energy Conservation. The Conservation Foundation, Washington, D.C., 1973.

Citizen Action Guide to Energy Conservation, Citizens' Advisory Committee on Environmental Quality. U.S. Government Printing Office, Washington, D.C., 1973.

Residential space heating is treated by Moyers in reference 1, and by Hirst and Moyers, "Potential for Energy Conservation Through Increased Efficiency of Use," in reference 8, p. 155.

Transportation energy use is discussed by Hirst in reference 2, by Hirst and Moyers in reference 8, p. 155, and in the following articles by Hirst.

Eric Hirst

"Energy Consumption for Transportation in the U.S.," ORNL-NSF-EP-15. Oak Ridge National Laboratory, Oak Ridge, Tennessee, March 1972.

"Energy Intensiveness of Passenger and Freight Transportation Modes, 1950–1970," ORNL-NSF-EP-44. Oak Ridge National Laboratory, Oak Ridge, Tennessee, April 1973.

"Transportation Energy Use and Conservation Potential," Science and Public Affairs, *Bulletin of the Atomic Scientists,* Vol. XXIX, No. 9, p. 36.

John R. Pierce

"The Fuel Consumption of Automobiles," *Scientific American* **232,** No. 1, p. 34.

9 Conclusions and Biased Opinions

"What do you paint, when you paint a wall?"
  Said John D.'s grandson Nelson.
"Do you paint just anything there at all?
"Will there be any doves, or a tree in fall?
"Or a hunting scene, like an English hall?"
  "I paint what I see," said Rivera.

E. B. White, "I Paint What I See" *

9.0   Introduction

The preceding chapters have given considerable information about many aspects of energy. Though the author's biases have crept in here and there, the presentation has been primarily factual. It would be inappropriate to leave the matter there, however; some interpretation of the information is in order. By its very nature, an interpretation cannot be objective in the way that facts are objective. At this point, then, the requirement of objectivity is relaxed and interpretations of and opinions on some of the information presented in the earlier chapters is given.

9.1   Energy Is Too Cheap!

In Chapter 6 we saw many ways by which some of the environmental damages caused by energy use can be reduced, if additional money is expended. Strip-mined land can often be reclaimed at a cost between 5 and 10 percent of the price of the coal removed. Aquatic thermal pollution can be eliminated by the use of dry cooling towers, at a cost of perhaps 15 percent of the price of the

---

* First stanza of "I Paint What I See" from *The Fox of Peapack* by E. B. White. Copyright 1933 by E. B. White. Originally appeared in *The New Yorker,* and reprinted by permission of Harper & Row, Publishers, Inc.

electricity produced. Sulfur can be removed from oil for $0.75/bbl, 15 percent of the price of the oil. In the same vein, solar energy, which is free from most environmental problems, can be used if we are willing to pay enough for it. It seems reasonable that we should increase the price of energy and reduce many of the associated environmental damages with the money generated thereby (internalize the externalities, to use the economists' jargon).

But, it is protested, modern industrial society is based on inexpensive energy. It is legitimate to ask how high the price of energy can rise (relative to the price of materials and labor) before the basic structure of the economy must change. If energy becomes too expensive, we would be forced to give up factory-produced commodities and return to handmade goods, give up tractors and return to farm animals, and so on. We will attempt to estimate this "switchover price."

There is no question that the price of energy in the United States has been very inexpensive (at least through 1973). Energy's 5-percent share of the GNP, based on its price, greatly understates the contribution that energy makes to our well-being and to production. Clearly we could pay considerably more than 5 percent of the national income for energy and still maintain a highly industrialized society. Also, the average consumer uses 7 percent of his personal income for energy (home heating fuel, electricity, gasoline), and he would pay a larger amount rather than change his life-style.

As noted in Chapter 7, the price of energy in England is about twice that in the United States, whereas the GNP per capita (and hence income per capita) is just above half that of the United States. This suggests that relative to labor, energy is four times more expensive in England than in the United States. Since they use only half as much per capita, they spend twice as large a fraction of their national and personal income on energy as we do in the United States. Since England is a modern industrial country, its very existence is proof that a four-fold increase in the price of energy relative to labor is possible. Considering western Europe as a whole, the conclusion is reinforced. This highly industrialized region uses about one-third the per-capita energy of the United States and pays at least twice as much (per Btu) for it. The cost of a Btu as a fraction of the region's income is at least four times that of the United States.

Taking one-third of the U.S. per-capita energy use as a conservative figure for the minimum energy requirement of a modern industrial society, and taking 20 percent of the national income as a conservative figure for the maximum fraction that could be allotted to pay for energy, we conclude that the U.S. price of energy (relative to labor) could rise a factor of 10 (but probably not much more) without causing a reversion back to a nonindustrial society. Clearly, internalizing energy's environmental external costs at price increases of less than a factor of 2 is something the United States can afford and should do. (Note that the factor of 10 increase in energy price per Btu is relative to the price of labor; some part of any large energy price increase would entail a decrease in real wages, i.e., in the material standard of living. With a factor of 3 reduction in per-capita energy use, GNP per capita would fall perhaps a factor of 2.)

The discussion of energy conservation and energy growth reduction in Chapter 8 did not consider the motivational aspect. It is evident that we waste energy so much because we value it so little. A significant increase in the price of energy would supply the motivation for the energy conservation steps discussed in Chapter 8. Energy is inexpensive in part because externalities are not included, but also in part because government policies subsidize the energy industries. Depletion allowances are a form of subsidy to the mineral extraction industries, the oil industry being a flagrant example. Federal dam-building projects subsidize hydroelectric energy. Tax deductions for oil royalties paid to foreign governments subsidize imported petroleum. Government research and development subsidizes nuclear energy. These government policies which subsidize the energy industries and thereby keep the price of energy low are poor policies from energy conservation and energy-growth-slowing points of view, and should be terminated. One can argue that energy should be priced *above* its "natural" value (e.g., by means of a tax) as a way of controlling energy growth. My own view is that pricing alone may not be sufficient to control energy growth. However, we should at least eliminate the subsidies and internalize the externalities so that energy is not priced *below* its "natural" value, i.e., so that economics helps rather than hurts.

In discussing price, it is necessary also to consider rate structure. As mentioned in Chapter 7, electricity and natural gas are sold at prices that decrease as the amount purchased increases. The last kWh or 100 cu ft, respectively, used in a given month is priced lower than the first, and the rate structure for industrial users is lower than that for residential users. This pricing structure encourages energy growth (its intended purpose) and discourages energy conservation (because you're always saving the low-priced energy rather than the high-priced energy). An "inverted rate structure," by which the price per kWh or 100 cu ft increases as the amount purchased increases, would give more incentive to conserve energy.

The logic behind a decreasing rate structure for electricity dates from the era when, as a result of technological improvements and economies of scale, newer and larger generating facilities could produce electricity more cheaply per kWh than older, smaller facilities. A pricing structure that encouraged growth resulted in construction of new facilities and thus lower prices per kWh for everyone. That era has ended. Diseconomies of scale have set in—as one moves toward a small number of very large generating plants, more reserve capacity (an extra cost) is required to cover the possibility of failure of one big plant. Increased costs of labor, materials, and capital have resulted in increased construction costs for new facilities. Fuel shortages (substantially caused by growth in energy use) have led to higher fuel prices. As a result of all this, we are now in an era where growth in electrical energy use is being accompanied by increases in electricity prices. This argues for flattening the rate structure, or perhaps even inverting it.

A consideration of external (environmental) costs certainly calls for an inverted rate structure. Environmental damage is not linear with insult; it stays small until a threshold is reached and then rises rapidly. The last Btu of heat into a river or last pound of SO_2 into the air causes more damage than the first. In more general terms, doubling the size of a power plant more than doubles the environmental damage it does. Alternatively, it more than doubles the cost of abatement equipment necessary to maintain a given level of damage, since abate-

ment costs also rise faster than linearly with, for example, percentage fly-a
removal. Although the utilities themselves have argued that the conventional rate
structure is not an example of the small user subsidizing the large user, I have
my doubts. Again, as with overall price, I would argue that the rate structure
should be flattened, or maybe even inverted, at least to the extent that payment
for the first kWh is not subsidizing the last kWh and environmental costs are
charged more heavily against the last kWh than the first.

Whenever anyone suggests raising the price of energy, someone asks, "What
about the poor?" My own feeling is that energy policy is not the proper vehicle
for solving the problem of poverty. However, in the case of the rate-structure
change suggested, the poor will benefit. Indeed I am surprised that the present
rate structures have not been more heavily criticized as examples of the poor
(who use less electricity and therefore are on the high-rate portion of the scale)
subsidizing the rich (who use more electricity and therefore are on the low-rate
portion of the scale).

9.2   Energy Growth Must Eventually Stop

Various limits to energy growth were mentioned at different points in the preceding chapters, principally in Chapters 4 and 6. We collect them here and translate them into a time at which the limit is reached, using a 20-year doubling time
for energy use (a reasonable value for the present and the recent past). It should
be realized that most of the limits are not abrupt (no problems before the limit
is reached, catastrophe when it is reached) but rather involve a general worsen-
ing of conditions which (in my estimation) will become severe at the point chosen
as the limit.

1. **Global thermal pollution** (Section 6.5). Man-made energy inputs of 1 percent of solar heating would warm the earth's surface by 1.5°C, and in all probability would cause major climatic changes. Present energy use is 0.005 percent of
solar, a factor of 200, or 7.5 doublings below the limit. With a 20-year doubling
time, the limit would be reached in 150 years.

2. **Regional thermal pollution** (Section 6.5). Man-made energy inputs of 5 percent of solar heating, averaged over a large region, would probably change convection patterns and cause significant changes in climate. For the eastern United
States and central western Europe, an order of magnitude increase (taking 70
years at present growth rates) would reach this limit.

3. **Carbon dioxide pollution** (Section 6.20). A doubling of atmospheric CO_2
would cause, via an enhanced greenhouse effect, a 2°C increase in the earth's
average surface temperature. A temperature increase as large as this would undoubtedly cause major climatic changes. Assuming that half of the CO_2 produced
by fossil-fuel combustion remains in the atmosphere, then the limit (doubling of
atmosphere CO_2) corresponds to combustion of 1400 billion tons of carbon. If
human energy needs are supplied principally by fossil fuels, this limit would be
reached in 65 years.

4. Fossil fuel supply (Section 4.4). World coal estimates of 7.5×10^{12} tons could supply all energy needs (with a 20-year doubling time) for 110 years. Note that the carbon dioxide pollution limit (item 3 above) is a factor of 5 more severe. Other fossil fuels (oil, natural gas) will be exhausted much sooner. The estimates of oil and natural gas given in Section 4.4 could supply all world energy needs for 48 years. Since oil and particularly natural gas have higher hydrogen-to-carbon ratios than coal, these fuels contribute less CO_2 per unit energy, and so their supply may well run out before CO_2 pollution reaches a limit.

5. Global sulfur dioxide problems (Section 6.19). Regions in western Europe already have freshwater aquatic problems (low pH) due to the presence of SO_2. Damage to soils is anticipated in 40 years. The region now affected has about 15 times the world average energy use per unit area for land surfaces. We take as a limit a world average SO_2 level of half that of the region now affected. Assuming no sulfur removal techniques are instituted and the same mixture of fuels is burned, this limit will be reached in 3 doublings, or 60 years. (On a regional level, western Europe is already at the limit.)

6. Deliberate consequences of energy use (Section 6.25). Our planet is modified by the intended uses of energy as well as by the side effects. Roads, houses, and factories are built, minerals are extracted, forests are harvested, etc. At some level, this intentional use becomes, of itself, too much, and the normal functioning of nature is greatly upset. I take as a worldwide limit that level of energy use equal to photosynthetic energy use. For a regional limit, I take 5 times photosynthetic energy. Current worldwide energy use is a factor of 10 less than land-based photosynthesis, and will equal it in 65 years. Eastern United States and western Europe are within a factor of 3 of the regional limit, and will reach it in 30 years.

7. Land availability. There is a limit to the fraction of the earth's land surface that can be tied up in the production and transport of energy. I take 5 percent as the limit. (For comparison, currently about 10 percent of the earth's land surface is cropland.) A crude upper limit of the amount of land affected by coal extraction can be obtained as follows. Assume all mineable coal is in a seam one meter thick and has a density of 1.35. Then one acre yields 5000 tons, and the estimated world reserves of 7.5×10^{12} tons occupy 1.5×10^9 acres, or 4 percent of the earth's land surface. This estimate will err on the low side because land adjacent to that from which coal is extracted is also affected. It will err on the high side because coal is frequently removed in amounts exceeding 5000 tons per acre. Assuming these roughly cancel, the limit placed by land availability is about the same as that placed by supply. It is hoped that by proper coal extraction techniques, this land availability limit can be effectively removed. Electrical transmission lines (Section 6.24) currently occupy 4 million acres in the United States and are projected to occupy 7 million acres (0.3 percent of the U.S. land surface) by 1990. Four more doublings are required for this to reach the 5-percent limit. It is difficult to convert this to a time interval because technical improvements (higher voltage lines) have been increasing the amount of power transmitted per power line. Also, electrical energy use is currently growing twice as fast as total energy use, but this must cease to be the case once electrical energy constitutes the major fraction of total energy. My conclusion is that, except as it is a component of item 6 above, bulk land availability need not be one of the earlier limiting factors. (This does not mean that use of land is unimportant—care should certainly be taken to minimize adverse effects, and in reaching my conclusion I am assuming that this will be done.) Energy growth impact on *specialized* land (agricultural land, coastal land, and wilderness) is more difficult to analyze, and I draw no conclusions, for example, as to what limit might be placed on offshore oil production by availability of coastal land.

   8. Aquatic thermal pollution (Chapter 6, Part B). In the United States and western Europe, problems already exist with once-through cooling. Assume evaporative cooling towers are used instead. Then if all electrical generating waste heat went to evaporate fresh water, a 20-fold increase from present levels would cause a consumptive use of 5 percent of U.S. river flow, which we take as a limit. Assuming electrical energy use doubles every 10 years until it is half of all energy use, and doubles every 20 years thereafter, the United States will reach this limit in 65 years. This limit can be circumvented by dry cooling towers or ocean cooling. The limit on ocean cooling is not clear to me, but is probably quite large. The limit on dry towers is climate modification (items 1 and 2 above).

   9. Urban air pollution (Section 6.18). The EPA has estimated the economic damages from urban air pollution at $16 billion per year. Comparing this figure to the wholesale price of energy (e.g., to a utility company), 60×10^{15} Btu/yr $\times$ 0.25 $/million Btu = $15 billion/yr, we see that the damages are comparable to the price. This implies that in urban areas, an energy limit has already been reached. Further energy increases require improved abatement techniques.

Comments

a) The above list of limits may well be incomplete. A deliberate omission is global particulate air pollution and its potential effects on climate. It is not evident to me how to obtain a numerical limit for this category.

b) No limits for nuclear energy (radioactivity) have been given. In principle (if everything goes according to plans), all radioactivity will be contained and there will be no problems. With the troubles due solely to accidents, it is difficult to derive limits. In expressing this view, I assume that ^{85}Kr and tritium will in the future be captured at the fuel reprocessing plants, and that the release to the environment of other radioactive materials can be further reduced if needs be. (These appear to be reasonable assumptions.) The fact that no limit is given for nuclear energy does not mean that it is harmless—see Section 6.16.

c) Some of the effects discussed are additive, so that the actual limit should be taken to be lower than that given for either one separately. Items 1 and 3 are in this category, as are items 5, 6, and 7.

d) If one "goes nuclear," or switches to fusion energy, then many of the limits disappear. Items 1, 2, 6, and 8 remain.

e) One of the tighter limits is item 6, deliberate consequences of energy use.

It suggests that world energy growth of a factor of 10 is undesirable, and that energy growth in the industrialized countries of a factor of 2 to 3 is undesirable. (For the United States as a whole, a factor of 2 sets human energy use equal to photosynthesis.) Admittedly, this particular limit is rather subjective and certainly not generally accepted. Indeed it is not very often even considered. It merits more attention.

9.3   Why Do We Need More Energy?

With energy growth substantially exceeding population growth both in the United States and worldwide, and with the damages caused by energy use becoming increasingly more severe, it is appropriate to ask, "Is that growth really necessary?" Many people have suggested that it is. We briefly look at their reasons.

One argument is, "We need energy growth for the poor." Here one must distinguish between poor nations (namely, nonindustrial nations) and poor individuals within rich nations (namely, industrial nations). Indeed if the nonindustrial nations are to become industrialized, then it is necessary for their per-capita energy consumption to rise, say to one-fourth or one-third that of the United States. While it is desirable from the point of view of an individual living in an industrialized nation that substantial portions of the globe remain undeveloped and nonindustrialized, it is unreasonable to expect that these regions should be entire nations or that any nation should forego the very considerable and very real benefits of industrialization. However, the need is for energy growth in the poor nations. This reasoning offers no justification for energy growth in the industrialized nations in general, and certainly none in the United States. Yet on an absolute scale, most of the world's increases in energy use are occurring in industrialized nations, not in developing nations.

When discussing poor individuals within rich nations (e.g., the poor in the United States), we must distinguish between their relative and absolute poverty. I believe it is the former that is important. It takes more money to live a tolerable life in a rich society than in a poor society. The individual's expectations are higher, the pressures (advertising and social) to buy higher-priced items are stronger, and the availability of lower-priced items is less. The usual argument in terms of energy for the poor is that increased energy use will mean increased per-capita GNP, and some small fraction of this will trickle down to the poor and reduce their absolute poverty. If absolute poverty is what counts, this is a rather inefficient approach. If relative poverty is what matters, this approach accomplishes nothing—increasing energy use or per-capita GNP will not guarantee a redistribution of wealth. As I stated earlier, energy policy is not the proper vehicle for solving the problems of poverty. Turning things around, assuming some more appropriate vehicle *does* solve the problem of poverty, what are the energy-policy and energy-growth implications? They are quite small. Suppose the economic level of the lower 20 percent of the population were raised to that of the average; this would require a less-than-20-percent increase in GNP, and similarly less-than-20-percent increase in energy use.

A second argument is, "We need energy growth because of population growth." In the United States, one-third of recent energy growth can be attributed directly to population growth. Worldwide, the figure is closer to one-half. Clearly some energy growth is required as long as there is a growing population. The question thus becomes rephrased, "Why do we need more people?" We don't! I have heard no good reason why we need a population increase, either worldwide or in the United States. I have heard numerous reasons why we would be hurt by such increases, and why we would be better off with lower populations.

A third argument is, "We need energy growth for environmental improvement." Since this argument is usually put forward by utility companies and not by people whose primary concern is environmental improvement, it should be viewed with some suspicion. Nevertheless it does merit careful examination. Indeed some solutions to some environmental problems do require the use of more energy. However, alternative solutions to these problems result in energy savings. Let's consider a few examples. First, eliminating aquatic thermal pollution with dry cooling towers (Section 6.9) raises the condenser temperature of steam-electric generating plants, lowers plant efficiency, and increases the energy input required to generate a fixed amount of electricity. Alternatively, using the waste heat for residential heating or some other beneficial purpose (Section 8.3) lowers the total energy requirement. Second, reducing automotive urban air pollution by emission control devices on automobiles lowers gasoline mileage and increases energy use. Alternatively, solving the urban air-pollution problem by a switch to mass transit in dense urban areas increases passenger miles per gallon and reduces energy use. Third, environmentally sound handling of solid waste will undoubtedly result in reduced energy use. If it is incinerated and the energy therefrom used, the solid waste will reduce the use of conventional fuels. If it is recycled, the products produced (paper, metals) will require less energy than if they were produced from virgin material (Section 8.3). Indeed, energy will be required to run the recycling plant, but energy will be saved at the pulp mill, aluminum mill, etc. Fourth, water pollution is an often-cited example of the need for more energy. Wastewater treatment plants require energy, and generally speaking the more effective the treatment is, the more energy is required. Yet even here there are energy-saving options. Natural gas can be made from sewage, thus reducing the use of "natural" natural gas. Fertilizer can also be made, thus saving the energy used in making fertilizer by conventional means. Fifth, land use is an area where energy is the problem, not the solution. Any answer to suburban sprawl, to proliferating highways, and to urban land abuse will generally result in use of less energy, not more.

From the preceding examples, we see that the orthodox "engineering approach" to most problems is to use more energy. This is quite natural, since energy has been so inexpensive, and applies to dealing with problems of all sorts, not just environmental ones. The engineering approach has been accused of being the *cause* of many environmental problems, because it tends to define its goals too narrowly and restrict its vision to only a small part of the system. Note that the alternative solutions that save energy invariably involve considering a larger system (e.g., electric generation plus home heating) or involve human behavioral changes (e.g., switching from autos to mass transit).

Let us take the pessimistic (realistic?) view that many environmental probems will be solved in an energy-intensive way. How much energy is required? The energy penalty in using dry cooling towers as compared to using once-through cooling is perhaps 5 percent (depending on climate), corresponding to a smaller increment than two years normal growth. The fuel penalty for automobile emission control is in the neighborhood of 5 to 10 percent, comparable with the penalty for a car air conditioner or automatic transmission, and considerably less than the difference in fuel consumption between full-size and subcompact cars. A recent study[1] found that all environmental standards now set to go into effect before the end of the decade will require only 3.5-percent increase in total energy use.

From the preceding it is apparent that large energy increases are not required to solve our environmental problems. Certainly only a very small part of energy growth in the recent past can be charged against environmental improvement.

A fourth argument favoring increased use of energy is, "We need energy growth (economic growth) to maintain full employment." It is beyond the scope of this book to discuss whether present economic systems require economic growth in order to maintain full employment. It is entirely possible that they do, since they evolved during an era when both economic growth and energy were universally accepted as "good things" (as indeed they then were). The argument, however, is backward. The social and economic systems are means to various ends, not ends in themselves. If consideration of natural laws suggests that further energy growth is undesirable and if the present socioeconomic system requires energy growth, then the system must be modified. It is no longer appropriate. The socioeconomic system cannot be used as an argument for energy growth.

In dismissing the argument I do not mean to dismiss the issue. If there is a conflict between a desirable end (i.e., greatly reduced energy growth) and a re-quirement of the present socioeconomic system (i.e., rapid energy growth), then it is very important to be aware of the fact. We must begin to consider how best to modify the system so as to preserve its desirable features but eliminate the undesirable ones.

A fifth argument is, "We need energy growth so the fat cats can get fatter." Now we're approaching the truth! At any rate, that's what past energy growth has been due to. But let's be sure we understand who the "fat cats" are. They are those living in the "rich" nations, i.e., the industrialized nations—western Europe, Japan, the Soviet Union, and (heading the list) the United States. Within these countries, they are the upper two-thirds of the population as far as per-capita income is concerned (not the upper 10 percent—the very rich are few enough in number that their consumption is not significantly important). In other words, the "fat cats" are you and me, and increased energy use is re-quired if we are going to have progressively more material things.

I'll defer to Section 9.5 the discussion of whether it is good for us to get fatter. Here we should only note that the good things related to energy increase with energy per capita, and the bad things related to energy increase with energy per square mile. The obvious moral is to keep the capita per square mile low, i.e., to keep a low population density. I do not mean its distribution; it is in fact beneficial to cluster population, to have high density in some small areas and very low density over most of the rest. I mean average population density, people per million square miles, people per planet.

9.4 Suppose We Ignore the Limits

Most of the limits mentioned in Section 9.2 pertain to the smooth functioning of natural ecosystems, of the ecosphere, of nature. Suppose we ignore the limits and allow nature to break down. It is easily within the scope of the imagination, and conceivably within the realms of possibility, that humans can dispense with nature and let technology take over all its roles. A view akin to this is expressed by Alvin Weinberg, "It is by now common to observe that with energy and readily available common rocks, one can in principle create the material basis for a tolerable life." [2]

The entire extent of human development has been one of "improving upon nature." Agriculture and animal husbandry have replaced hunting and gathering; housing and clothing provide protection from the elements, and so on. Having made all these technological advances, it is easy to conclude that with a little more effort, man could "go it alone." Man's success in such nonnatural environments as cities and spacecraft reinforces this impression. It is important to review the long list of "goods and services" furnished us by nature before deciding that we can do everything by means of technology, and to fully appreciate what doing without nature would entail. A small number of examples are listed below.

a) Natural phenomena give rise to a mean temperature on the earth's surface of 14°C (57°F), a quite tolerable temperature for humans, about 15°F lower than their preference. The fluctuations in surface temperature are very often less than ±40°F, a fairly narrow range. (Consider for comparison temperatures on the moon or on other planets.) Our present space-heating and air-conditioning efforts should be viewed as "fine-tuning" nature, not as dispensing with it.

b) The hydrological cycle "irrigates" a large fraction of the earth's land surface. Human irrigation projects are minor by comparison. In conjunction with biological processes, the hydrological cycle suplies pure water for domestic and industrial purposes. The human role in water purification is small compared to the natural role.

c) The composition of the atmosphere is maintained and air pollutants are removed by natural processes.

d) Human food supply can be viewed as a cooperative venture between nature and man. Water supply, nutrient supply, and pest control are all items that man can regulate to some extent. However nature certainly plays the dominant role here. Man is a long way from being able to create foodstuffs from energy, carbon dioxide, and water in a factory on a scale that could feed the earth's present population.

e) Other items include flood control (through vegetative cover), insect control (indeed, control of the numbers of all species), and lumber production.

As man takes over for nature, human energy requirements must increase, resulting in further interference with natural processes. For example, consider increasing the crop yield from some farmland by intensive human intervention (fertilizer and pesticides). Environmental damage is caused by the process of extracting and utilizing the energy to manufacture the fertilizer and pesticides. In addition, environmental damage is caused by fertilizer runoff and pesticide residue. The environmental damage caused in these various ways may reduce the "goods and services" delivered by nature in some other area, necessitating further human intervention. For example, (1) the fertilizer runoff could destroy a water supply, requiring a human purification system. (2) The pesticide use could cause a breakdown of natural pest control in forest areas adjacent to the farmland, causing loss of timber (3) Thermal pollution from the power plant that generates electricity used in fertilizer manufacture could reduce the fish yield in the ecosystems affected by the river used for cooling, necessitating increased food production somewhere else. The important point is that the feedback is positive. At some level of human energy use, of human intervention in nature, the positive feedback of the mechanism just described will overpower the negative feedback mechanisms that keep nature in balance. This level is the "point of no return." Once it is passed there is no turning back; one must dispense entirely with nature and "progress" to an artificial world.

The point of no return is of course related to the various limits given in Section 9.2. It is the lowest unavoidable limit given there, which turns out to be item 6 (deliberate consequences of energy use). I am suggesting that once human energy use is comparable to photosynthetic energy on a global scale, we risk damaging nature to the point that we must take over all her functions and considerably increase human energy use in the process. (Of course, the point of no return is not determined exclusively by energy use. World population, pesticide levels, water pollution, and many other factors also influence it. However, even if all these other factors were optimal, the point of no return would still exist and would be reached by too high a level of human energy use.)

The view of a world drifting from a natural state toward an artificial state has been graphically described by J. J. Fremlin.[3] Even though his concern was with population growth, his description is equally applicable to energy growth. Assuming that each limit posed by population growth (energy growth) is confronted and solved by technology, he describes a world becoming more artificial, step by step, until at last the insurmountable problem of heat dissipation is encountered. By then the earth is a hermetically sealed single building 2000 stories high, and world population is $\sim 10^{17}$ people, 120 per square meter. There is a flavor of science fiction to the essay, but the message is clear—if we circumvent the limits placed on population growth (energy growth) by technology, the natural world will disappear. Further, if we do not make a conscious decision to stop population growth (energy growth), then the natural drift of things will lead us toward that hermetically sealed, 2000-story world.

The limits mentioned in Section 9.2 thus confront us with a choice. We can either stop energy growth or "go artificial" and dispense with nature. It is not clear that the second alternative will work—our technology may not be up to it. More to the point, however, is that even if it does work it will result in a world far inferior to the one we live in now. In the absence of an explicit decision to halt energy growth, we will by default "go artificial." This is certainly the direction we are heading. The decision must be made in the next few decades and implemented within a century.

9.5   Consequences of Zero Energy Growth

Population

With no energy growth, population growth must also stop or else the energy per capita (and hence the material standard of living) will decrease. As a practical matter, population growth should stop before energy growth stops. There are compelling reasons for stopping population growth now—food supply via ecologically sound agriculture and land use are two. Indeed there is considerably more consensus in favor of stopping population growth than there is for stopping energy growth.

Gross National Product

With no energy growth, there can be no growth in GNP of the conventional sort, i.e., growth involving manipulating materials. There can be shifts and changes in what is produced, perhaps resulting in increased value of goods produced for the same amount of materials manipulation. Perhaps there can be growth in those components of GNP that require very little energy such as services, entertainment, information, handcrafts, and the arts.

Would a "no-growth" society be pleasant or even bearable? It's hard to say. Concerning the lack of growth in per-capita GNP, it must be realized that the relationship between personal well-being or happiness and material goods is not linear, but behaves as shown schematically in Fig. 9.1. For example, the second 1000 calories per day of foodstuffs benefit an individual less than the first 1000. In turn, the third 1000 are less beneficial than the second, and the fourth 1000 calories per day are apt to be harmful rather than beneficial. The case is similar with living space, space heating, etc. For some items (e.g., foodstuffs), an optimum can be readily established. For others, all that can be said is that with increased amounts of the good in question, a situation of diminishing returns sets in. Since happiness is a psychological variable, its dependence on material goods can be varied by advertising and other social pressures. It is my belief that in the United States and most other industrialized societies, most individuals are beyond the steeply rising portion of the curve, and their happiness can be increased more by nonmaterial than material things.

Considerable concern has been expressed that a no-growth society would involve excessive restrictions on personal freedoms. The extent to which this is the case would depend on the value at which the population density levels out—

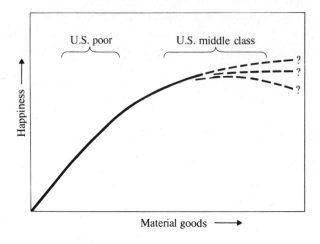

Fig. 9.1 Dependence of personal well-being (happiness) on amount of material goods. (Intended as a qualitative representation only.)

the higher the population density, the less the personal freedoms. It seems almost certain that an artificial world would necessarily involve considerably more restrictions on personal freedom than a no-growth society would.

In the same manner, a no-growth society would appear to severely limit opportunities for innovation and change. This is probably true. However, the increasing interdependencies of an artificial world would also place severe restrictions on permissible changes.

In summary, while the no-growth society does not look particularly attractive, it looks far better than any alternative.

How about the transition period from our present growth-oriented society to a no-growth society? It is bound to be difficult. Our social and economic systems evolved during the growth-oriented period of the last few hundred years. It would be quite surprising if they were at all suitable for a no-growth society. Yet it's not clear how to achieve the necessary modifications, or even what modifications would be desirable. On a more practical level, many industries and occupations now depend on growth (e.g., construction of homes). There would be unemployment problems to be coped with in making the transition. Also, as population growth comes to a halt, the age distribution will change as it approaches an equilibrium. As the number of very young decreases and the number of very old increases relative to the group in the middle, various adjustments would be required.

9.6   Possible Scenarios for the Future

To estimate ultimate levels of energy use, we must consider both population and per-capita energy use. World population is unlikely to level off at anything below twice its present level, and it could well rise as high as four times the present level. U.S. population might level off as low as 1.5 times the present population, or it could go as high as twice the present level. Assuming a fully industrialized world, per-capita energy use could not be much below half the current U.S. level (most of the industrialized countries today are near half the U.S. level), and it might go as high as twice the present level. U.S. per-capita energy use could conceivably fall slightly, say to two-thirds of its present level, as a result of conservation measures and reduced emphasis on material goods (idealistic dreaming, of course, put in to obtain a lower limit). More likely it will rise, easily to twice its present level.

The above figures suggest that world energy use will level off at 5 to 40 times present world energy use. With present energy growth rates, the leveling off would occur in 40 to 100 years. The five-fold increase in world energy use is below the limits listed in Section 9.2; the forty-fold increase is above most of them. U.S. energy use might level off at 1 to 4 times present U.S. energy use. The four-fold increase would have U.S. human energy use at twice U.S. photosynthetic activity, probably rather harmful for a region as large as the United States.

The choice of energy sources for both the near-term and the more distant future is large, and each source has its advantages and disadvantages. *Coal* has the advantage of a very large supply, but it is not as convenient to transport or use as the fluid fossil fuels, and its extraction and use involve severe environmental problems. (Some of the disadvantages of coal can be circumvented by converting it to oil or gas.) The situations with *oil* and *gas* are mirror images of coal. Easy to transport and use, and causing considerably less environmental damage than coal, they are in short supply, particularly in the United States (relative to U.S. energy needs). There is, however, a strong possibility that as the supplies of oil and gas dwindle, the environmental damages associated with their extraction and transportation will increase. All the fossil fuels share the long-term limitation of CO_2 production; therefore they cannot be the principal energy source for the long haul.

Fission energy bypasses most of the environmental damages of the fossil fuels, but it creates the possibility for some new ones through accidents involving radioactivity. Without breeder reactors, the supply of nuclear fuel is large but not unlimited. With breeders, the fuel supply becomes essentially unlimited, but the likelihood of accidents probably increases. *Fusion energy* (as imagined) should be environmentally very attractive; the only certain environmental problem is thermal pollution. Fuel supply is unlimited. The major disadvantage of fusion energy is that it is not now a working system, and it is not known when it will work, or even if it will work.

Solar energy is environmentally benign. Because it is a continuous source, supply is not an issue. Its technological status is somewhat confusing. At a price it will work for any purpose; for home heating and hot-water heating it is economically competitive today in some regions. Whether technology can sufficiently reduce the cost of solar energy for other uses (e.g., electric power generation) is unknown. *Geothermal energy* is currently fraught with great uncertainties as to its usable reserves; estimates vary by two orders of magnitude, from negligible to quite significant. Other sources (e.g., hydropower and tidal power) will not play a major role on the world energy scene, though they may have local or regional importance.

One projection of U.S. energy sources to the year 2000 is given in Fig. 7.4. Throughout this period, oil and gas supply the bulk of the energy, though fission energy is rapidly becoming more important. Beyond the year 2000, the relative importance of fossil fuels would continue to decrease, and the role of nuclear energy would continue to increase, though at a slower rate. Breeder reactors would contribute to the energy picture by the year 1990, and fusion perhaps by 2010.

The above projections might be labeled as the *1970 conventional wisdom scenario*. It has several defects and there are alternative scenarios (with different defects). The defect receiving the most recent attention, as a result of balance-of-payment problems and the 1973–1974 Arab oil embargo, is the dependence of the 1970 conventional wisdom scenario on imports of oil and gas. To minimize this, the *1974 conventional wisdom scenario* suggests we use more coal, initially in solid form (relaxing environmental regulations to permit this), but as soon as possible (10 years?) as oil and gas from coal. The *antinuclear scenario* suggests heavy reliance on fossil fuels until the year 2000, and strenuous research activity on geothermal energy, solar energy, and fusion energy so that fusion or solar energy can take the major burden after the year 2000. Use of fission energy (the breeder in particular) can be largely bypassed. Those particularly concerned about the dangers of fission energy favor this course. The *pro-nuclear scenario* suggests heavy use of light-water reactors as soon as possible, followed by heavy use of breeders as soon as possible (say by 1990) followed by fusion energy as the principle source (say by 2020).

One cannot evaluate the merits or even the feasibility of the various scenarios without first making some assumptions about energy growth. For example, if our high estimate of U.S. energy use (four-fold increase) materializes, and does so at present growth rates (in 40 years, two doubling times), then the antinuclear scenario would not be viable. Problems of fossil fuel supply would be severe and the environmental ill effects acute. For such a growth in energy use, the pro-nuclear scenario seems to be the best option, the risk of nuclear accident being the price paid for continued energy growth. On the other hand, if U.S. energy use levels out not far above its present value (our low estimate), then the antinuclear scenario would be very viable, and would at first be fairly close to maintaining the status quo.

9.7   The Energy Crisis

The term "energy crisis" is much in use these days, though it has not appeared often in this book. I have avoided using the term because it conjures up quite different meanings to different people. Many of the meanings of the term have been discussed, however. To some, energy crisis means a shortage of energy supply (see Chapter 4). To others, it is a shortage of the wherewithall (e.g., generating capacity or refining capacity) for putting the existing supply to use. (This aspect has not been discussed in this book.) Still others consider the crisis to be the environmental damage caused by energy use, and others, conversely, consider it as the limitations placed on energy use by environmental regulations (see Chapter 6). A final point of view is that the crisis is not one of supply but of demand, demand growing exponentially with no hint of leveling off (see Chapter 7). Indeed all of these interpretations are a part of the energy crisis. It is important to understand that one should not consider only one and exclude the others.

Of late, the appropriateness of the word "crisis" as a description of our energy problems has been questioned. When used in a medical sense, "crisis" refers to a brief period of time, whose outcome determines whether the patient survives or succumbs. On the time scale that we are accustomed to using, then, our energy problems are not a crisis because the relevant time period is not hours or days, but decades. From the perspective of an individual living through this period, there will be no conspicuous turning point, no obvious recovery, but only a continuous series of problems and difficulties. On the time scale of human history, however, the decades of the 1970's and 1980's indeed are a moment of crisis when the crucial decision about limiting energy growth must be made. The outcome of this brief period will determine the fate of patient earth.

Questions and Problems

9.1
Describe the changes you would expect in the U.S. economy if the price of energy (relative to materials and labor) rose by a factor of 3.

9.2
Make a complete list of all government subsidies to the energy industry. Include items that the energy industry denies are subsidies, but that its "opponents" assert in fact are. Give a quantitative description of each entry in your list. Estimate the magnitude of the total annual subsidy; express it as a percentage of the dollar value of annual energy production. (This is a term project.)

9.3
a) A typical family of four has a monthly residential electricity use in the neighborhood of 750 kWh, exclusive of electrical space heating. Determine their monthly electricity bill using the rate schedule given in Table 7.7. Assume through conservation measures that they reduce their monthly electricity use by 30 percent. By what percentage will their bill drop? Assume through carelessness and conspicuous consumption their monthly electricity use doubles (to 1500 kWh per month). By what factor will their bill increase?

b) Devise an inverted rate structure that will encourage conservation and penalize waste. Note its effects in the above case. Give a justification of the inverted rate structure on grounds other than energy conservation.

9.4
Make your own estimates of the various limits to energy growth. Specifically, criticize the limits given in Section 9.2, modify the limits as you feel appropriate, and give any other limits you feel are important. Attempt a quantitative justification of all limits.

9.5

Do you agree that deliberate consequences of energy use (item 6 of Section 9.2) constitute a serious hazard to the environment? Discuss your answer, being quantitative wherever possible.

9.6

To what extent and for what purposes do you feel further energy growth is necessary and/or desirable?

9.7

Assume it is desired to "dispense with nature" and run the planet Earth with modern technology alone (alone the lines described in Section 9.4). List all the major functions now performed by nature that technology would have to take over.

9.8

Discuss what a no-growth society would be like.

9.9

Is there an optimum level of material goods per capita? How could one define such an optimum? How could one determine it? On which side of the optimum is the U.S. middle class today—too many or too few material goods? On which side of the optimum are you?

9.10

Criticize each of the scenarios of energy sources in the future given in Section 9.6. Propose and justify two scenarios of your own: (a) the most probable, and (b) the most desirable (yet realistic).

Notes

1

John Davidson and Marc Ross, "Energy Needs for Pollution Control During the Next Decade," Report to the Ford Foundation Energy Policy Project. To be published; see EPP under Related Reading.

2

A. M. Weinberg, "Prudence and Technology—A Technologist's Response to Predictions of Catastrophe." Keynote address, 1970 IEEE Nuclear Science Symposium, New York, November 4, 1970.

3

J. F. Fremlin, "How Many People Can the World Support?" *New Scientist,* No. 415, 1964, p. 285.

Related Reading

I am confident that my conclusions and biased opinions are correct, and I offer them as Revealed Truth. For perspective, however, you should be familiar with conclusions and biased opinions of others, covering a wide range of viewpoints. While the following references are presented for their interpretive content, some also contain useful factual information.

Neil Fabricant and Robert M. Hallman
Toward a Rational Power Policy—Energy, Politics, and Pollution. New York: George Braziller, 1971. A report by the New York City Environmental Protection Administration, discussing problems associated with electrical power.

John J. Murphy (ed.)
Energy and Public Policy—1972, A conference report. New York: The Conference Board, 1972.

Energy: The Need is NOW. New York: National Association of Manufacturers.

John C. Fisher
"Energy Crisis in Perspective," *Physics Today* **26**, No. 12, p. 40.

EPP
Exploring Energy Choices, A preliminary report of the Energy Policy Project of the Ford Foundation, Washington, D.C., 1974.

EPP: A Time to Choose, Final report of the Energy Policy Project of the Ford Foundation. Cambridge, Mass.: Ballinger Publishing Company, 1974. The EPP is a major study of the energy issues facing the United States. The EPP has an interdisciplinary staff coordinating the effort and also has commissioned two dozen studies of various aspects of these issues by independent groups. The EPP is funded by the Ford Foundation at a level of $4 million. Many of the special research studies are to be published in a series of books during the latter half of 1974 by Ballinger Publishing Company.

A. M. Weinberg and R. P. Hammond
"Limits to the Use of Energy," *American Scientist* **58**, No. 4, p. 312.

"Global Effects of Increased Use of Energy," in *Peaceful Uses of Atomic Energy,* Vol. I, Proceedings of the 4th International Conference on the Peaceful Uses of Atomic Energy, Geneva, September 1971. New York: United Nations, 1972, p. 171. The technological optimist's point of view. See also Note 2.

Outlook for Energy in the United States to 1985, Energy Economics Division, Chase Manhattan Bank, June 1972.

Sam H. Schurr (ed.)
Energy, Economic Growth, and the Environment, Papers presented at a forum conducted by Resources for the Future, Washington, D.C., April 1971. Baltimore: Johns Hopkins Press, 1972.

Alvin Kaufman and Thomas E. Browne
The Energy Crisis—An Overview, OER Report No. 12. An internal report prepared by the Office of Economic Research, New York State Department of Public Service, Albany, November 1972.

Science
Science **184**, No. 4134. This issue of *Science* is devoted completely to energy and contains both factual and interpretive articles.

David J. Rose
"Energy Policy in the U.S.," *Scientific American* **230**, No. 1, p. 20.

Foreign Policy Implications of the Energy Crisis, Hearings before the Subcommittee on Foreign Economic Policy of the Committee on Foreign Affairs, House of Representatives, 92nd Congress. Washington, D.C.: U.S. Government Printing Office, 1972.

Appendix A
A Short Course in Thermodynamics

Thermodynamics is the study of energy transformations. Historically it has been particularly concerned with transformations of heat energy into mechanical work, and of mechanical work into heat energy. The fundamentals of thermodynamics are contained in the First and Second Laws.

First Law of Thermodynamics

The First Law is a statement of energy conservation—the total energy of a closed system is a constant, neither increasing nor decreasing with time. For a system not gaining or losing matter but otherwise open, the First Law states that the change in the energy of the system is equal to the work done on the system plus the heat flowing into the system. If we denote the energy of the system by U, the work done by it by W, and the heat flow into it by Q, then the above statements of the First Law may be written:

$$\Delta U = 0 \qquad \text{(closed system)} \qquad\qquad\qquad (A.1)$$

$$\Delta U = Q - W \qquad \text{(system closed to material exchanges).} \qquad (A.2)$$

The term W includes mechanical work, but it also includes electrical work, magnetic work, etc., that is, flow of energy in an organized, nonmaterial form. Note both W and Q can be positive or negative—work can be done by or on the system; heat can flow into or out of the system. Recognizing that "work done on the system" and "heat flowing into the system" are both examples of energy flowing

244

into the system, a third, equivalent statement is: The change in the energy of a system is equal to the algebraic sum of all energy flows into the system. And a fourth statement is: Energy is neither created nor destroyed, but is transformed from one form to another and can move from place to place. The historical importance of the enunciation of the First Law is that it represented the recognition that heat is a form of energy.

Second Law of Thermodynamics

The Second Law places restrictions on possible energy transformations. The two classic (equivalent) statements are: (1) A transformation whose only final result is the extraction of heat Q_1 from some body at a temperature T_1 and the conversion of all of it to work $W = Q_1$ is impossible. (2) A transformation whose only final result is the transfer of heat Q from a body at a given temperature T_1 to a body at a higher temperature $T_2 > T_1$ is impossible.

The two statements can be proven to be equivalent by noting that a violation of one statement makes possible the violation of the other statement. Assume statement (1) were false. Then the work W obtained by extracting heat Q_1 at temperature T_1 could all be used to add heat to a body at temperature $T_2 > T_1$, for example by friction or by electrical resistance heating. The heat added would equal $W(= Q_1,)$, by the First Law, and the net effect of the combined process would be the transfer of heat Q_1 from a body at T_1 to a body at $T_2 > T_1$. But this is a violation of statement (2). Similarly, assume statement (2) were false. Then after transferring heat Q from a body at T_1 to a body at $T_2 > T_1$, we could operate a heat engine between these two bodies, allowing heat Q to flow out of the body at T_2, converting part of the heat to work W while the rest $Q' = Q - W$ flows to the body at T_1. The net result has been the removal of heat $Q - Q'$ from the body at T_1 and its complete conversion to work. But this is a violation of statement (1). Since the falsity of either statement implies the falsity of the other, they are either both true or both false. That is, they are equivalent.

Carnot Cycle

We know that it is possible to cause heat to flow from lower to higher temperatures *if work is done* (refrigerators) and that it is possible to obtain work by extracting heat from a body *as long as some heat flows to a body at a lower temperature* (steam engines). Since the Second Law says that the italicized quantities above cannot be zero, we now investigate what limits can be placed on them by considering the *Carnot cycle*. Consider two heat reservoirs (bodies so large that they can absorb or give up heat with negligible change in temperature) at temperatures T_1 and T_2 ($T_2 > T_1$). Further, consider a *heat engine* (a gas inside a cylinder, for example) operating between these two reservoirs. The engine operates by going through the following reversible four-step cycle: (1) At a temperature T_2 and in contact with the higher temperature heat reservoir, the engine absorbs an amount of heat Q_2 (e.g., by allowing the gas in the cylinder to expand isothermally against a movable piston). (2) Insulated from both reservoirs, the working medium of the engine drops in temperature from T_2 to T_1 (e.g., by a further, adiabatic expansion). (3) At a temperature T_1 and in contact with the lower temperature heat reservoir, the engine gives up an amount of heat Q_1 (e.g., by isothermal compression of the gas). (4) Insulated from both reservoirs, the working medium of the engine rises in temperature from T_1 to T_2 (e.g., by further, adiabatic compression). At the end of these four steps, the engine is back where it started and is ready to repeat the process, i.e., it has gone through a cycle. Figure A.1 illustrates the cycle for an engine consisting of gas in a cylinder with a movable piston. All conclusions to be obtained are independent of the nature of the engine, however. All that matters is that the cycle consists of two isothermal steps and two adiabatic steps, *and that the process is reversible.* The requirement of reversibility, i.e., that the system must at all times be in equilibrium, can be met in principle by going through the cycle very slowly (infinitely slowly). During any or all of the four steps, the engine may be doing external work or having work done on it. The net work done by the engine, by the First Law, is $W = Q_2 - Q_1$. The cycle just described is known as the Carnot cycle, and an engine working in such a cycle is known as a Carnot engine. The requirement of reversibility has the consequence that the cycle can be run through in the opposite sense, with W, Q_1, and Q_2 having the same magnitude but flowing in the

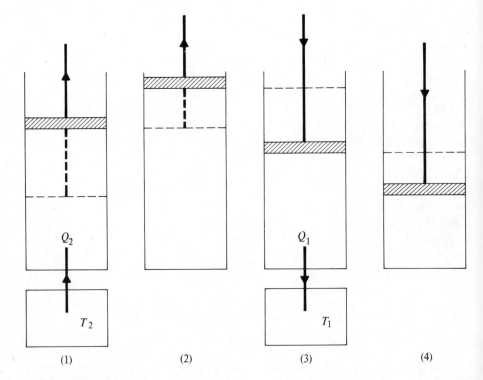

Fig. A.1 Carnot engine consisting of a gas in a cylinder with a movable piston. The four steps of the cycle are shown.

opposite direction. Running this way, the engine operates as a refrigerator, moving heat from the colder to the hotter reservoir and having work done on it.

The efficiency η of a heat engine is appropriately defined as that fraction of the heat absorbed that is converted to work, i.e.,

$$\eta \equiv \frac{W}{Q_2} = \frac{Q_2 - Q_1}{Q_2} = 1 - \frac{Q_1}{Q_2}. \tag{A.3}$$

Statement (1) of the Second Law says that η must be less than 1. We shall now prove that (a) the efficiency of any heat engine operating between two temperatures cannot exceed that of a reversible engine operating between the same two temperatures (i.e., a Carnot engine), and (b) all reversible engines operating between the same two temperatures have the same efficiency.

Proof of (a). Assume the contrary, namely, that there exists a heat engine with efficiency $\eta > \eta_{\text{carnot}}$. Allow this engine to run and let it drive the reversible engine as a refrigerator. Then the engine can furnish work W to drive the refrigerator by removing $Q_2 = W/\eta$ units of heat from the high-temperature reservoir. The refrigerator will deliver $Q_{2c} = W/\eta_{\text{carnot}}$ units of heat to the high-temperature reservoir. The net result of this combined cycle will be transferring

$$Q_{2c} - Q_2 = \frac{W}{\dfrac{1}{\eta_{\text{carnot}}} - \dfrac{1}{\eta}}$$

units of heat from T_1 to T_2, a positive quantity if $\eta > \eta_{\text{carnot}}$. This violates statement (2) of the Second Law; therefore $\eta \lessgtr \eta_{\text{carnot}}$.

Proof of (b). This follows immediately from (a). If there were two reversible engines with different efficiencies, then one would be larger than the other. That is, one would have an efficiency larger than some Carnot engine, contrary to (a).

Since the efficiency of a reversible engine is the maximum possible efficiency, there is considerable interest in determining it. All reversible engines operating between T_2 and T_1 have the same efficiency. Therefore, this efficiency (and therefore Q_1/Q_2) must depend *only on T_2 and T_1*.

$$\frac{Q_1}{Q_2} = f(T_1, T_2). \tag{A.4}$$

Now consider a third temperature T_0. Imagine a pair of Carnot cycles, between T_2 and T_0 and between T_0 and T_1. Arrange them so that one adds to T_0 the same amount of heat as the other removes:

$$\frac{Q'_2}{Q_0} = f(T_2, T_0), \frac{Q'_1}{Q_0} = f(T_1, T_0); \qquad \frac{Q'_1}{Q'_2} = \frac{f(T_1, T_0)}{f(T_2, T_0)}.$$

But the net effect of this pair of cycles is just that of a single Carnot cycle operating between T_1 and T_2, so

$$\frac{Q_1}{Q_2} = \frac{Q'_1}{Q'_2}; \qquad f(T_1, T_2) = \frac{f(T_1, T_0)}{f(T_2, T_0)}.$$

Since T_0 was arbitrary, we may choose some reference temperature, writing $g(T_1) = f(T_1, T_0)$, and obtaining

$$\frac{Q_1}{Q_2} = \frac{g(T_1)}{g(T_2)}. \qquad (A.5)$$

One can determine $g(T)$ by considering a particular Carnot engine, specifically an engine using an ideal gas as a working medium. In this way it can be shown (although we shall not do so) that $g(T) = \text{constant} \times T$, and therefore

$$\frac{Q_1}{Q_2} = \frac{T_1}{T_2}, \qquad (A.6)$$

where the temperature is in an absolute scale.

The Carnot efficiency follows from the above expression:

$$\eta_{\text{carnot}} = 1 - \frac{T_1}{T_2}. \qquad (A.7)$$

Since T_1 is always greater than zero, η_{carnot} is always less than 1, in agreement with statement (1) of the Second Law. As T_1 approaches T_2, η_{carnot} approaches zero, again in agreement with statement (1). The larger the fractional difference between T_1 and T_2, the larger the possible efficiency. As T_1 approaches absolute zero, the efficiency approaches unity.

One can also consider the efficiency of refrigerators, Carnot and otherwise, which is conveniently defined as the ratio of the heat removed from the cooler body to the work expended in removing it, that is;

$$\epsilon = \frac{Q_1}{W}. \qquad (A.8)$$

As for heat engines, one can prove that no refrigerator can have an efficiency larger than that of a Carnot refrigerator. This limiting efficiency is

$$\epsilon_{\text{carnot}} = \frac{T_1}{T_2 - T_1}. \qquad (A.9)$$

This "efficiency" can be larger than one, and becomes very large when the temperature difference through which the heat is to be transferred $(T_2 - T_1)$ is a small fraction of T_1. However, as long as $T_2 - T_1$ is not zero, ϵ_{carnot} is not infinite, in accord with statement (2) of the Second Law.

Entropy

The concept of entropy enables the Second Law to be stated in a concise, mathematical form and can give one a deeper understanding of it. Unfortunately, most people acquire an intuitive feel for the meaning of entropy rather slowly. Here we will first give a formal definition of entropy and restate the Second Law using it. Then a qualitative discussion of the meaning will be given.

For a Carnot cycle we saw that

$$\frac{Q_2}{T_2} - \frac{Q_1}{T_1} = 0.$$

Now generalize by considering a system B taken through an arbitrary cycle, during which it exchanges heat with any number n of heat reservoirs of different temperatures, and may also do work or have work done on it. Denote by T_i and Q_i the temperature of the ith reservoir and the heat flowing from the ith reservoir to the system, respectively. (If heat flows from the system to the reservoir to the system, then Q_i will be negative.) We shall show that

$$\sum_{i=1}^{n} \frac{Q_i}{T_i} \leq 0 \tag{A.10}$$

and that the equality holds if the cycle is reversible.

Proof: After B has been taken through the cycle, return the heat Q_i to each of the n heat reservoirs by transferring heat from an additional heat reservoir at temperature T_0 using Carnot cycles. At the end of all this, B is back to where it started and the n heat reservoirs are back to where they started. Heat Q_0 will have flowed out of the additional reservoir and (by the First Law) gone into work done by the Carnot engines or by B. By the Second Law this cannot occur, that is, Q_0 cannot be greater than zero. To evaluate Q_0, note that in its cycle with the ith reservoir, the additional reservoir gives up heat $Q_{0,i} = T_0(Q_i/T_i)$.

$$Q_0 = \sum_{i=1}^{n} Q_{0,i} = \sum_{i=1}^{n} (T_0) \frac{Q_i}{T_i} = (T_0) \sum_{i=1}^{n} \frac{Q_i}{T_i}. \tag{A.11}$$

Since T_0 is a positive number, then

$$Q_0 \leq 0 \Rightarrow \sum_{i=1}^{n} \frac{Q_i}{T_i} \leq 0 \qquad \text{QED.}$$

If the cycle that B goes through is reversible, then the entire process can be run through backwards, with the result that the negative of the above sum must also be nonpositive ($x \leq 0$ and $-x \leq 0 \Rightarrow x = 0$). For a reversible cycle, the equality in Eq. A.10 holds.

We have considered heat exchanges with a finite but arbitrarily large number of reservoirs n. This is readily generalized to an infinite number of reservoirs, i.e., the sum converts to an integral:

$$\oint \frac{dQ}{T} \leq 0. \tag{A.12}$$

The circle on the integral sign signifies that the integral is taken around a cycle (a closed path). T is the temperature of the reservoir exchanging heat dQ—not necessarily the temperature of the system at that time. For a reversible process,

$$\oint \frac{dQ}{T} = 0 \qquad \text{(reversible processes).} \tag{A.13}$$

Now consider two states X and Y of some system B. (X is shorthand for pressure, temperature, etc., namely, everything necessary to specify the state of B.) Define S by

$$S(Y) - S(X) = \int_{X}^{Y} \frac{dQ}{T} \qquad \text{(reversible path),} \tag{A.14}$$

where the integral is evaluated along any reversible path between X and Y, i.e., B is changed reversibly from state X to state Y. Under most conditions, an integral depends on the choice of path as well as on the end points. However, because of Eq. A.13, the integral in Eq. A.14 is *independent of path*. Any reversible path from X to Y will give the same result. (*Proof:* Consider two reversible paths (1) and (2) between X and Y. As shown in Fig. A.2, these two paths constitute a closed curve. One can integrate from X to Y along path (1), and then back from Y to X along path (2). The sum must be zero, and therefore the integral from X to Y along path (1) equals the negative of the integral from Y to X along path (2), i.e., it equals the integral from X to Y along path (2) QED.) The quantity S defined by Eq. A.14 is (surprise!) the *entropy*. The entropy is a function of the state of a system, but *not* of how the system arrived at that state. Equation A.14 defines the entropy only up to an additive constant, and is therefore adequate for determining entropy changes, but not for determining the actual value. For our purposes this is sufficient.

The First Law in differential form is written

$$dU = dQ - dW, \tag{A.15}$$

where dU is the change in energy of some system, dQ is the heat energy flowing into the system, and dW is the work done by the system. The term dW can be written in terms of the state variables of the system, e.g., PdV for expansion or compression of a gas ($P =$ pressure; $V =$ volume). For reversible changes $dQ = TdS$, which is similar in form.

Equation A.12 can be shown to be equivalent to either of our original statements of the Second Law. Using it and the definition of entropy, one can write

$$\int_X^Y \frac{dQ}{T} \gtrless S(Y) - S(X). \tag{A.16}$$

The closed curve used to obtain this result goes from X to Y along the path of interest, i.e., the path actually followed by the system, and then returns from

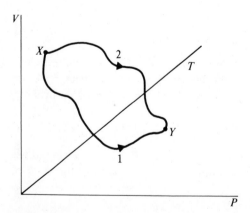

Figure A.2

Y to X along any reversible path. The definition of entropy is used to replace the integral on the return path. Note that since the path followed by the system is *from* X to Y, the system is at Y at a later time than it is at X. In differential form, Eq. A.16 is written

$$dQ \leqslant TdS. \tag{A.17}$$

A case of particular interest is that of a closed system. Then dQ must vanish and Eq. A.16 becomes

$$\Delta S \geqslant 0 \qquad \text{(closed system).} \tag{A.18}$$

This leads to yet another statement of the Second Law—the entropy of a closed system does not decrease. Left to itself, a system will proceed from a state of low entropy to a state of high entropy.

As an example, consider a system consisting of two blocks of metal, one at a temperature T_1 and the other at a higher temperature T_2. If these blocks are put in thermal contact with each other, heat will flow from the hotter one to the cooler one until they are both at the same temperature T_3 somewhere in between T_1 and T_2. We calculate the entropy change of the system by bringing both blocks to T_3 *reversibly*. Since heat is flowing out of the high-temperature block, its entropy decreases. On the other hand, heat is flowing into the low-temperature block, so its entropy increases. Since the same amount of heat flows for both blocks,

$$\int \frac{dQ}{T}$$

for the low-temperature block is larger than for the high-temperature block. Therefore, the entropy increase of the low-temperature block is larger than the entropy decrease of the high-temperature block, and the net change is an increase in entropy for the system, in accord with Eq. A.18.

The formal definition of entropy gives little insight into its physical significance. However, the example given (as well as many others) suggests that entropy is a measure of the order or disorder of a system. A highly ordered system has a small value of entropy; a highly disordered system has a large value of entropy. In the example of the two metal blocks, the situation with most of the heat energy stored in one block was more ordered than the situation with the heat energy spread uniformly over the blocks. A better example is the phase changes of water. The molecules of ice at $0°C$ are neatly arranged in a crystalline lattice, whereas the molecules of liquid water at $0°C$ are in random motion over roughly the same volume. Clearly, ice is a much more ordered system than liquid water and should therefore have considerably lower entropy. Indeed, melting ice requires adding a substantial amount of heat and therefore a substantial entropy increase. In comparing liquid water with water vapor, the vapor occupies a much larger volume (typically 1000 times larger) than the same amount of water. For this reason the vapor is much more disordered. (Liquid water can be neatly set in the corner of a room in a pail, whereas the same amount of water vapor is spread all over the room.) Again, one expects that the entropy of water vapor should be much higher than that of liquid water, and therefore boiling water should require adding a substantial amount of heat. This is indeed the case.

The Second Law statement that $\Delta S \geqq 0$ (closed system) can thus be phrased: Left to itself, a system proceeds from a state of order to a state of disorder.

Our definition of entropy has been a thermodynamic, macroscopic definition. One can also give a statistical mechanical, microscopic definition:

$$S = k \ln W, \tag{A.19}$$

where $k = 1.38 \times 10^{-23}$ joules/deg is the Boltzmann constant and W is the *thermodynamic probability* of the state of the system. Roughly speaking, W is the number of different ways the molecules of the system can be distributed to give the same macroscopic state. While one cannot calculate W without resorting to quantum mechanics, one can calculate the ratio of W's for two different states of a system. For example, consider n gas molecules, spread over a volume V_1 (thermodynamic probability W_1) or over a volume V_2 (thermodynamic probability W_2). The number of different ways each molecule can be distributed is proportional to the volume it is to be distributed over, implying $W \propto V^n$. Then

$$\frac{W_2}{W_1} = \left(\frac{V_2}{V_1}\right)^n \quad \text{and} \quad S_2 - S_1 = (k \ln W_2) - (k \ln W_1) = k \ln \frac{W_2}{W_1} = n\, k \ln \frac{V_2}{V_1}.$$

Clearly there is a connection between order/disorder and thermodynamic probability. Highly ordered states have small thermodynamic probability—there are relatively few ways the molecules can be distributed to yield those states. Highly disordered states have large thermodynamic probability—the molecules can be distributed many ways to yield the states. (Recall the ice/water/water vapor situation.)

Thermodynamic Potentials

For a purely mechanical system, (a) all the energy of the system is available to do work, and (b) the system tends toward a state of minimum potential energy, which is a stable equilibrium. For thermodynamic systems the situation is changed. There are functions with the dimensions of energy that satisfy conditions (a) and (b). However, they are *not* the total energy and potential energy, but rather the various *thermodynamic potentials*.

Define the following thermodynamic potentials:

$$H = U + PV \qquad \text{(enthalpy)} \tag{A.20}$$

$$F = U - TS \qquad \text{(Helmholtz function)} \tag{A.21}$$

$$G = U - TS + PV \qquad \text{(Gibbs function)} \tag{A.22}$$

Each is a function of the state of the system to which it refers. Their properties follow from the First and Second Laws:

$$dU = dQ - dW \tag{A.15$'$}$$

$$dQ \lessgtr T dS. \tag{A.17$'$}$$

We will investigate changes in each of the thermodynamic potentials as a system changes from state (1) to state (2).

Enthalpy. Consider first the enthalpy. Consider changes taking place at constant pressure $(P_2 = P_1)$. Then

$$\Delta H = H_2 - H_1 = Q - W + P(V_2 - V_1) \tag{A.23}$$

For many systems the only work done is by expansion $(dW = PdV)$. Under such conditions $W = P(V_2 - V_1)$ and $\Delta H = Q$. The change in enthalpy is equal to the heat flowing into the system. Consider a system consisting of some fuel and the oxygen necessary to burn it. Combustion usually takes place at constant pressure and does work only by expansion. Then the heat released by the fuel-oxygen system (available to drive a boiler, for example) is equal to the decrease in enthalpy of the system.

Helmholtz function. Next, consider changes in the Helmholtz function due to system changes taking place at constant temperature $(T_2 = T_1)$.

$$\Delta F = F_2 - F_1 = -W + Q - T(S_2 - S_1) \lesseqgtr -W \tag{A.24}$$

or

$$W \lesseqgtr -\Delta F.$$

The Helmholtz function, therefore, is the energy that is available to do work at some (constant) temperature. Suppose a system is dynamically isolated from its surroundings (held at fixed volume, etc., so that it can neither do work on its surroundings nor have work done on it) but kept in thermal contact with a reservoir at constant temperature. Then $W = 0$ and $\Delta F \lesssim 0$. Such a system tends toward a state of minimum Helmholtz function, which is a stable equilibrium.

Gibbs function. Finally consider changes in the Gibbs function due to system changes taking place at constant temperature $(T_2 = T_1)$ and pressure $(P_2 = P_1)$.

$$\Delta G = G_2 - G_1 = Q - T(S_2 - S_1) - W + P(V_2 - V_1) \lesseqgtr -W + P(V_2 - V_1). \tag{A.25}$$

Frequently one is interested in work other than that done by expansion. For example, in a battery it is the electrical work that is of interest. Define such "non-PdV work" by $dL = dW - PdV$. Then under conditions of constant temperature and pressure, $L \lesssim -\Delta G$. The Gibbs function, therefore, is the energy that is available to do non-PdV work at constant temperature and pressure. For a system isolated so that it is unable to do non-PdV work on its surroundings, but held at constant temperature and pressure, $L = 0$ and $\Delta G \lesssim 0$. Such a system tends toward a state of minimum Gibbs function, which is a stable equilibrium.

As seen above, F and G are under appropriate circumstances measures of the energy available to do desired work. For this reason, the name *free energy* has been used for both functions, causing some confusion. For processes proceeding at constant pressure and involving only small volume changes, the distinction between F and G is unimportant. In order to keep the main text understandable to a more general audience, we use the term "free energy" for both F and G, specifying which is meant if the difference is important.

The conditions of constant temperature and pressure, under which G has interesting properties, apply to most biological chemistry—reactions proceed at atmospheric pressure and ambient temperature and cause negligible change in either. Further, the reaction usually does no non-PdV work on its surroundings, so the reaction proceeds in such a way as to minimize the Gibbs function.

The Gibbs function for a dilute solution containing K different solutes in a single solvent, with N_i moles of solute, $i = 1, K,$ and N_0 moles of solvent, can be written

$$G = \sum_{i=0}^{K} N_i \left(f_i(T) + Pv_i(T) \right) + RT \sum_{i=1}^{K} N_i \ln \frac{N_i}{N_0}. \tag{A.26}$$

$R = 8.3$ joules/deg is the gas constant, and $f_i(T)$ and $v_i(T)$ are functions of temperature characteristic of the ith substance in solution. This equation may be rewritten

$$G = \sum_{i=0}^{K} G_i = \sum_{i=0}^{K} N_i \left(g_i(T,P) + RT \ln \frac{N_i}{N_0} \right). \tag{A.27}$$

The Gibbs function per mole of solute,

$$\frac{G_i}{N_i} = g_i(T,P) + RT \ln \frac{N_i}{N_0},$$

contains a term independent of concentration and a term depending only on concentration and independent of the nature of the solute i. Chemical reactions will proceed so as to minimize G. They will proceed away from substances with large g_i and toward substances with small g_i. They will not proceed all the way, however, because of the concentration term. As the concentration becomes high, the concentration term increases, compensating for the small g_i, and the reaction stops. Similarly, as the concentration becomes low, the concentration term decreases, compensating for the large g_i and the reaction stops.

Appendix B
A Short Course in
Nuclear Physics

Nuclear Forces

Atomic nuclei are composed of neutrons and protons, collectively called *nucleons*. The nuclear force between two nucleons (two neutrons, two protons, or a neutron and a proton) is very complicated in all its detail. The gross features are as follows.

a) When the separation distance between the nucleons is very small, the force is very strongly repulsive. (Like marbles, nucleons resist being pushed too close together.)

b) At a slightly larger separation distance, the force becomes strongly attractive.

c) As the separation distance is further increased, the force quickly drops to zero.

An idealized plot of potential energy *vs.* two-nucleon separation distance is given in Fig. B.1. An overall characterization of the nuclear force is that it is attractive, strong, and short-range. For the proton-proton case, the coulomb force must be added to the nuclear force. It is repulsive, not too strong, and long-range (the potential falls as $1/r$). At short distances, the nuclear force dominates, whereas at larger distances, the coulomb force is more important.

Binding Energy

Consider some nucleus ${}_{Z}^{A}X_{N}$, composed of N neutrons, Z protons and, therefore, $A = N + Z$ nucleons. The energy necessary to dissociate the nucleus into A

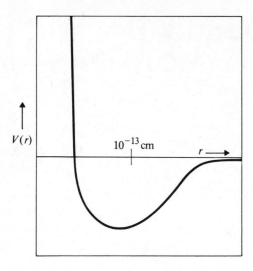

Fig. B.1 Potential energy $V(r)$ *vs.* two-nucleon separation distance r.

separate nucleons, divided by the number of nucleons A, is called the *binding energy per nucleon* of the nucleus X. Our interest is in how the binding energy per nucleon depends on the size of the nucleus, i.e., on A. The binding of a nucleon inside the nucleus will be constant, independent of the size of the nucleus. The nucleon experiences an attractive force from all the nearby nucleons, but *only* from the nearby nucleons (because of the short-range character of the nuclear force). A nucleon on the surface of the nucleus not surrounded by other nucleons experiences an attractive force from fewer nucleons, and is therefore less tightly bound. For a very low-A nucleus, most of the nucleons are on the surface, and are therefore loosely bound. Conversely, for a high-A nucleus, most nucleons are inside and therefore more tightly bound. A curve of binding energy per nucleon *vs.* A (Fig. B.2) shows these features. As A increases, the binding energy rises to about 8 MeV and then remains constant.

The existence of the coulomb force alters the above picture in a crucial way. Because the coulomb repulsion is long-range, it is effective over the entire nucleus, and therefore becomes increasingly important as Z becomes larger. Its effect is to lower the binding energy at large A, as shown in Fig. B.2. Maximum binding thus occurs in the medium-A region.

Neutron-to-Proton Ratio

The nature of the nuclear force, in conjunction with the Pauli exclusion principle, is such that the most stable nucleus consisting of A nucleons has equal numbers of protons and neutrons. Nuclei with different ratios exist but are not as tightly bound, and in some cases are unstable. Again, the existence of the coulomb force

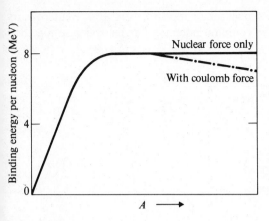

Figure B.2

alters this picture at high A; nuclei with a larger fraction of neutrons are more stable because with fewer protons, the coulomb repulsion is reduced. Figure B.3 shows the line in the N–Z plane where the stability is greatest, both for the case of nuclear force only and for the actual case of nuclear plus coulomb forces.

Fission and Fusion

Energetics. From the binding energy/nucleon *vs.* A curve (Fig. B.2), we see that nuclei are most tightly bound in the medium-A region. One can therefore *release energy* by moving from either end of the curve toward the middle. Taking a very heavy nucleus and breaking it into two medium-A pieces will release about 1 MeV of energy per nucleon. For example:

$$^{236}U \rightarrow {}^{96}X + {}^{140}Y + \sim 200 \text{ MeV}. \tag{B.1}$$

Alternatively, taking two very light nuclei and sticking them together to make a heavier nucleus also releases energy. For example:

$$d + d \rightarrow t + p + 4 \text{ MeV}, \tag{B.2}$$

and

$$d + t \rightarrow {}^4He + n + 17.6 \text{ MeV}. \tag{B.3}$$

The process of breaking a large nucleus into smaller nuclei is called *fission,* and the process of combining small nuclei to make a larger nucleus is called *fusion*. Fission and fusion are the two sources of nuclear energy.

Mechanism of fusion. Since the reaction $d + d \rightarrow t + p$ releases energy, the right-hand side is more stable than the left-hand side, and the reaction should tend to proceed spontaneously, i.e., the deuterons should attract each other. At short range they will; however, at long range there is coulomb repulsion. The potential *vs.* deuteron separation distance is shown schematically in Fig. B.4. Deuterons require an energy E_1 to overcome the coulomb potential barrier and fuse together.

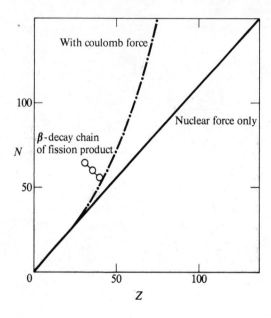

Figure B.3

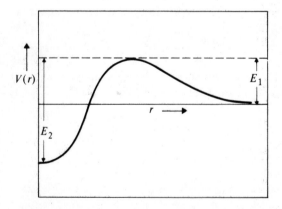

Figure B.4

Given E_1, they will fuse and release more energy E_2. A net energy $E_2 - E_1$ is thus extracted. Schemes for utilizing fusion as an energy source are discussed in Section 5.3.

Mechanism of fission. The situation with fission is similar to that of fusion. A large nucleus would "like" to become two smaller ones, becoming more stable and releasing kinetic energy in the process. However, to do this it must pass through an intermediate state of *higher* energy, represented by the middle diagram of Fig. B.5. As shown in the potential drawing, there is a barrier to overcome, this time a "surface-energy" barrier. (In the middle diagram, more of the nucleons are near or at the nuclear surface, so that the average binding energy is less, i.e., the potential is higher.) If an energy E_c is given to the nucleus, it overcomes the barrier and releases more energy E_2. A net energy $E_2 - E_c$ is thus extracted. (The quantum-mechanically sophisticated reader will realize that the nucleus can also penetrate the barrier without requiring that the energy E_c be added. This process is called *spontaneous fission;* it does occur, although at a lower rate.) Schemes for utilizing fission as an energy source are discussed in Section 5.2.

Radioactivity

Not all nuclei are stable. Some can decay into other states: $A \rightarrow B + x$. Some examples are:

$$^{238}_{92}\text{U}_{146} \rightarrow {}^{234}_{90}\text{Th}_{144} + \alpha \qquad (\alpha\text{-decay})$$

$$^{35}_{16}\text{S}_{19} \rightarrow {}^{35}_{17}\text{Cl}_{18} + e^- + \nu \qquad (\beta\text{-decay})$$

$$^{137}_{56}\text{Ba}^*_{81} \rightarrow {}^{137}_{56}\text{Ba}_{81} + \gamma \qquad (\gamma\text{-emission})$$

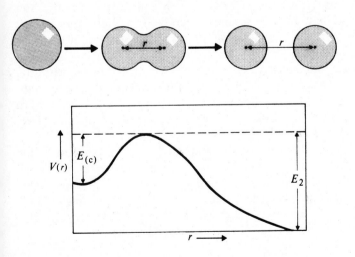

Figure B.5

In α-decay, the emitted particle (α) consists of two protons and two neutrons; in β-decay, an electron (e^-) and a neutrino (ν) are emitted; the γ-ray is a quantum of electromagnetic energy like a photon or an x-ray. The asterisk in the third reaction indicates that the barium nucleus is initially in an excited state.

The decay of a radioactive nucleus $A \to B + x$ is a probabilistic phenomenon. That is to say one cannot predict *when* a decay will take place, but only state that for any time interval, there is a certain *probability* that a decay will take place. For short time intervals, it turns out that this probability is proportional to the number of nuclei A at the beginning of the time interval, i.e., the number of candidates for decay. Denote by $I_A(t)$ the number of nuclei of type A at time t. Then the number of decays between t and $t + \Delta t$ is $\lambda\, I_A(t)\, \Delta t$, where λ is a constant of proportionality whose significance we will see shortly. These decays subtract from I_A:

$$I_A(t + \Delta t) = I_A(t) - \lambda\, I_A(t)\, \Delta t \tag{B.4}$$

yielding

$$\frac{dI_A}{dt} = -\lambda\, I_A, \tag{B.5}$$

which is solved to give

$$I_A(t) = I_A(0)\, e^{-\lambda t}. \tag{B.6}$$

Letting $\tau = 1/\lambda$, we have

$$I_A(t) = I_A(0)\, e^{-t/\tau}. \tag{B.7}$$

The number of nuclei decay away exponentially, with a lifetime τ. That is, after τ seconds, only $1/e$ are left; after 2τ seconds only $(1/e)^2$ are left; after 3τ seconds only $(1/e)^3$ are left, etc. τ is the *mean-life* associated with the decay. One can also define the *half-life* $T = \tau \log_e(2) = 0.693\tau$. After one half-life, half of the nuclei have decayed and half have not; after two half-lives $1/4$ are left; after 3 half-lives $1/8$ are left; after n half-lives $(1/2)^n$ are left. The value of the half-life (or the mean-life) is a characteristic of the specific decay in question. Each radioactive species has its own half-life and the range covered extends from fractions of a second to millenia.

Often a decay product B is also unstable and will decay. For example, fission products are always "neutron rich," i.e., they lie to the left of the stability line in Fig. B.3. They will move toward the stability line by a series of β-decays. Some of the intermediate states may be excited states rather than ground states. These will decay by γ-emission to the ground state. As an example, the decay chain of fission product ^{87}Br is

$$^{87}\text{Br} \xrightarrow{\beta-,\ 56\ \text{sec}} {}^{87}\text{Kr}^* \xrightarrow{\gamma} {}^{87}\text{Kr} \xrightarrow{\beta-,\ 1.2\ \text{hrs}} {}^{87}\text{Rb} \xrightarrow{\beta-,\ 6\times10^{10}\ \text{yrs}} {}^{87}\text{Sr}\ (\text{stable}).$$

A few excited states (for example, $^{87}\text{Kr}^*$) can decay by neutron emission as well as by γ-emission (i.e., $^{87}\text{Kr}^* \to {}^{86}\text{Kr} + \text{n}$). This phenomenon occurs for only a few nuclei, but it plays a crucial role in controlled fission (see Section 5.2).

Interactions of Neutrons with Nuclei

Neutrons passing through matter will "bump into" the nuclei of the matter and can interact with these nuclei in a variety of ways. Three ways of particular importance are elastic scattering, inelastic scattering, and neutron capture.

Elastic scattering, denoted by the reaction formula

$$n + {}_Z^A X_N \to n + {}_Z^A X_N, \tag{B.8}$$

is a billiard-ball-like collision of the neutron with the nucleus. The neutron just "bounces off" and the nucleus recoils a little. The nucleus thus gains a little kinetic energy but no excitation energy, i.e., it remains in its ground state. By energy conservation, the kinetic energy gained by the nucleus is lost by the neutron. The heavier the nucleus, the smaller the energy transfer—see Problem 5.3. (If the process is viewed in the center of mass, neither particle gains or loses energy.)

Inelastic scattering is denoted by the reaction formula

$$n + {}_Z^A X_N \to n' + {}_Z^A X_N^*. \tag{B.9}$$

In this reaction the nucleus acquires internal energy as well as kinetic energy, i.e., it is raised to an excited state (indicated by the asterisk). (Just as with atoms, the internal energy of nuclei is not a continuous variable, but consists of a series of discrete levels or states. Under normal circumstances, a nucleus will be found in its lowest level, called the *ground state*. The first level above the ground state (*first excited state*) is typically a fraction of an MeV above the ground state. Inelastic scattering can raise a nucleus to its first excited state or to a higher state. As mentioned in the preceding section, nuclei usually decay from excited states by γ-emission, but in a few cases can decay in other ways.) Again by energy conservation, the energy gained by the nucleus is lost by the neutron. Because the nucleus gains excitation energy as well as kinetic energy, the neutron's energy loss is larger than in the case of elastic scattering. (If the process is viewed in the center of mass, both particles lose kinetic energy.)

Neutron capture is denoted by the reaction formula

$$n + {}_Z^A X_N \to {}_Z^{A+1} X_{N+1}^* \tag{B.10}$$

In this case, the colliding neutron is "gobbled up" by the nucleus, and a new nucleus is formed containing the same number of protons but one more neutron. Invariably it is formed in a highly excited state.

The probability of a neutron undergoing any of the above interactions is proportional to the amount of material traversed. The constant of proportionality is called the *cross section* and has the dimensions of area. Each interaction has its own cross section, which depends on the neutron energy and on the particular nucleus in question. Thus one speaks of the capture cross section for neutrons on ^{238}U, the elastic scattering cross section for neutrons on carbon, etc. One can think of a particular cross section as being the effective cross-sectional area of the nucleus for the interaction under consideration—if the neutron "hits" that area, the interaction takes place; if it misses, the interaction doesn't. Thus the probability of a neutron undergoing an interaction is proportional to the cross section for the interaction: large cross section means high probability, small cross section means low probability.

We note two general features of neutron cross sections. (1) Inelastic-scattering cross sections are zero for neutron energies below the energy of the first excited state of the nucleus in question. (The neutron doesn't have enough energy to excite the nucleus, so the reaction cannot go.) (2) There is a region of energy (typically 1 eV to 100 keV) where neutron-capture cross sections undergo rapid variations. In this region, there are several very narrow energy bands where the capture cross section is very large. Capture within these bands is referred to as *resonance capture*. The behavior typical of the resonance-capture region is shown schematically in Fig. B.6.

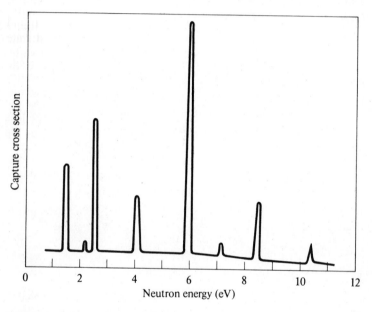

Fig. B.6 Schematic representation of energy dependence of neutron-capture cross section in the resonance capture region.

Appendix C
Electromagnetic Radiation

Radio waves, infrared radiation, visible light, ultraviolet light, x-rays, and gamma-rays are all examples of electromagnetic radiation. These various forms differ greatly in their frequency and wavelength, and therefore in the way they interact with matter. Because of this, we perceive them to be quite different from one another. Nevertheless, all forms are described by the same underlying physics, that of electromagnetic radiation. There are two "pictures" of electromagnetic radiation—wave and particle—which at first sight appear to contradict each other. Further consideration suggests that these two pictures are *complementary*, that there are circumstances in which one is more appropriate than the other, and that *both* are needed to give a complete description.

The *wave picture*, which evolved during the nineteenth century, views electromagnetic radiation as a wavelike "motion" of electrical and magnetic fields. As with water waves, these electromagnetic waves travel, transport energy, and are characterized by a wavelength, a frequency, and a velocity. A mathematical description of such a wave is

$$\vec{E}(\vec{r},t) = \vec{E}_0\, e^{i(\vec{K}\cdot\vec{r} - \omega t)} \tag{C.1}$$

This describes a plane-wave propagating in the direction of the vector $\vec{K}$ with a velocity

$$\omega/|\vec{K}|.$$

The frequency is $\nu = \omega/2\pi$; the wavelength is

$$\lambda = 2\pi/|\vec{K}|.$$

In terms of ν and λ, the velocity $v = \lambda\nu$. The velocity of an electromagnetic wave traveling in matter (e.g., glass, water, etc.) depends on frequency in a way that depends on the substance being traversed. However, the velocity of electromagnetic radiation traveling in free space (i.e., in vacuum) is a universal constant (usually denoted c) independent of frequency. In free space, $\lambda = c/\nu$, and so either wavelength or frequency is sufficient to specify the wave properties. The universal constant c has a value very close to 3×10^8 m/sec.

The *particle* or *quantum picture* of electromagnetic radiation evolved during the first half of the twentieth century in response to the wave theory's inability to describe certain interactions of electromagnetic radiation with matter. In several cases (notably the photoelectric effect, blackbody radiation, and Compton scattering), it appeared that energy was being exchanged between matter and radiation not in a continuous manner over space and time (as was expected from the wave picture), but rather discretely at an instant in time and localized in space. Electromagnetic energy appeared to come in little bundles rather than in waves. In the quantum picture, electromagnetic radiation of frequency ν is viewed as being made up of particles or quanta, each having an energy $h\nu$ and traveling (in free space) with a velocity c. The constant of proportionality between energy per quantum and frequency is known as Planck's constant, is denoted by h, and has the value 6.63×10^{-34} joule-sec.

As mentioned above, neither the wave picture nor the particle picture gives a complete view of electromagnetic radiation. One is more applicable than the other in a certain situation, and together they give a full picture. The various regions of the electromagnetic spectrum and the wavelengths, frequencies, and energies per quantum appropriate to each are shown in Fig. C.1.

Of particular concern to us in our treatment of energy is how electromagnetic radiation interacts with bulk matter. Electromagnetic radiation incident on bulk matter is reflected, transmitted, and absorbed. If we denote by r, t, and a the fractions of the electromagnetic energy that meet each fate, then conservation of energy implies $r + t + a = 1$. The quantities r, t, and a depend on the nature of the material, the frequency of the radiation, and the angle of incidence. For an opaque body, $t = 0$ (by definition) and $r + a = 1$.

In addition to the above-mentioned processes, a hot piece of matter will radiate, i.e., give off electromagnetic radiation when none is incident. Everyday examples of this include an electric iron, a burner on an electric stove, and an incandescent light bulb. From these and other examples we note that (1) the higher the temperature of the object, the more energy is given off, and (2) the higher the temperature of the object, the "bluer" the radiation, i.e., the higher the frequency and the shorter the wavelength. The connection between temperature and electromagnetic radiation can be qualitatively understood as follows. *Temperature* means thermal motion, or the random accelerations of atomic constituents of the hot body. But these atomic constituents, nuclei and electrons, are charged. Shaking charged particles around gives rise to electromagnetic radiation. The "harder the shaking" (the higher the T), the more radiation is emitted and the higher is the frequency of radiation.

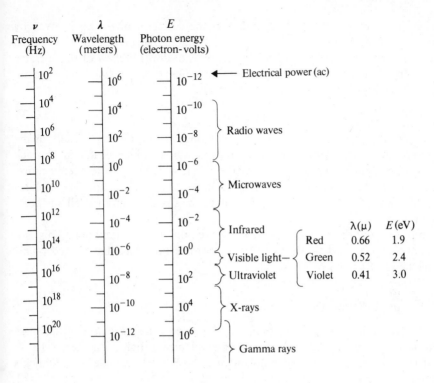

ν Frequency (Hz)	λ Wavelength (meters)	E Photon energy (electron-volts)				

Fig. C.1 Regions of the electromagnetic spectrum.

Stefan's law quantitatively relates the temperature of an object to the total radiant energy it emits. It was originally (1879) proposed as an empirical relation, but can now be derived theoretically. (We will not do so here.) The law states:

$$I = \sigma \epsilon T^4, \tag{C.2}$$

where T is the absolute temperature in °K, ϵ is the emissivity of the surface, a number between 0 and 1, which depends on the nature of the surface, $\sigma = 0.567 \times 10^{-11}$ joule cm^{-2} deg^{-4} sec^{-1} (Stefan-Boltzmann constant), and I is the radiant energy emitted (summed over all frequencies) per second per cm^2 of emitting surface. As an example, we calculate the energy radiated from a 100-cm^2 surface with emissivity of ½ at 400°K (slightly hotter than boiling water).

$$\text{Energy/sec} = 0.567 \times 10^{-11} \frac{\text{watts}}{\text{cm}^2 \text{ deg}^4} \frac{1}{2} (400)^4 \text{ deg}^4 \times 100 \text{ cm}^2 \approx 6 \text{ watts}.$$

If the temperature is raised from 400°K to 500°K, the radiated power increases by a factor of $(500°\text{K}/400°\text{K})^4$, that is, to 15 watts.

Our main interest will be the exchange of energy between bulk matter and an electromagnetic radiation field. For this, two aspects of the surface of the bulk matter concern us. (1) How much radiation the surface emits—this is given by Stefan's law, with the emissivity ϵ characterizing the surface. (2) How much radiation the surface absorbs—this is given by the *absorptivity* a of the surface, being the fraction of the incident radiation that is absorbed. Now it can be proven (Kirchhoff, 1860) that for any given surface,

$$\epsilon = a. \tag{C.3}$$

Very briefly, the proof proceeds as follows. Consider two objects initially at the same temperature enclosed in a perfectly reflective box, emitting radiation at each other. If a/ϵ for one is less than a/ϵ for the other, it will absorb less, emit more, and cool, while the other absorbs more, emits less, and warms up. This violates the second law of thermodynamics. Therefore, a/ϵ is the same for all surfaces. We can set $\epsilon = a$.

Consider now the frequency distribution of the emitted radiation. This is a famous problem, solved by Planck (1901), with a postulate that started quantum mechanics. Planck's result is:

$$I(\nu)\,d\nu = \frac{2\pi\nu^2}{c^2}\left(\frac{h\nu}{e^{h\nu/kT}-1}\right)\,d\nu, \tag{C.4}$$

where $I(\nu)\,d\nu$ is the energy with frequency between ν and $\nu + d\nu$ emitted from the surface per unit area per second, c is the velocity of light, k is 1.38×10^{-23} joule deg^{-1} (Boltzmann's constant), and h is 6.63×10^{-34} joule-sec (Planck's constant). This result was derived for a "black body," i.e., one for which $a = 1$ and, therefore, $\epsilon = 1$. It can be generalized to any surface by multiplying the right-hand side by ϵ, but now ϵ may depend on frequency, i.e., $\epsilon = \epsilon(\nu)$.

Often it is more convenient (or conventional) to use wavelength rather than frequency as the variable. Since the two are uniquely related, a mathematical mapping will take us from $I(\nu)\,d\nu$ to $\mathscr{J}(\lambda)\,d\lambda$, the energy per unit area per second emitted from a surface, with wavelength between λ and $\lambda + d\lambda$. Curves of $\mathscr{J}(\lambda)$ *vs.* λ are shown for a black surface ($\epsilon = 1$) at three different temperatures in Fig. C.2. As the temperature increases, the intensity increases and the peak of the intensity shifts to shorter wavelengths. For temperatures in the range of hundreds of degrees, most of the radiation is in the infrared region. At 5800°K, however, a major portion of the radiation comes in the region of visible light.

In the preceding discussion, we have been somewhat sloppy in our treatment of the frequency dependence of ϵ and a. In Stefan's law (Eq. C.2), the symbol ϵ represents $\epsilon(\nu)$ appropriately averaged over the frequencies of the emitted radiation, and might better be denoted $\bar{\epsilon}$. For determining what fraction of the incident radiation a surface will absorb, the relevant absorptivity a is $a(\nu)$ appropriately averaged over the frequencies of the incident radiation, better denoted $\bar{a}$. The Kirchhoff relationship (Eq. C.3) should be written $\epsilon(\nu) = a(\nu)$; absorptivity and emissivity need be equal only at the same frequency. If the frequency range of incident and emitted radiations are different, the averaging procedure will be different, so that $\bar{\epsilon}$ *need not* equal $\bar{a}$.

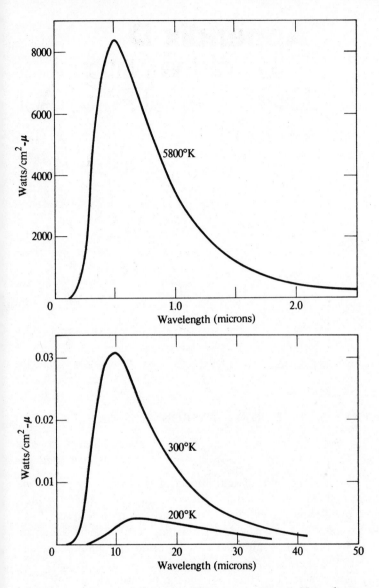

Fig. C.2 Black-body radiation at different temperatures. Note changes in scale.

Appendix D
Title I of the National Environmental Policy Act of 1969

DECLARATION OF NATIONAL ENVIRONMENTAL POLICY

SEC. 101

(a) The Congress, recognizing the profound impact of man's activity on the interrelations of all components of the natural environment, particularly the profound influences of population growth, high-density urbanization, industrial expansion, resource exploitation, and new and expanding technological advances and recognizing further the critical importance of restoring and maintaining environmental quality to the overall welfare and development of man, declares that it is the continuing policy of the Federal Government, in cooperation with State and local governments, and other concerned public and private organizations, to use all practicable means and measures, including financial and technical assistance, in a manner calculated to foster and promote the general welfare, to create and maintain conditions under which man and nature can exist in productive harmony, and fulfill the social, economic, and other requirements of present and future generations of Americans.

(b) In order to carry out the policy set forth in this Act, it is the continuing responsibility of the Federal Government to use all practicable means, consistent with other essential considerations of national policy, to improve and coordinate Federal plans, functions, programs, and resources to the end that the Nation may—

(1) fulfill the responsibilities of each generation as trustee of the environment for succeeding generations;

(2) assure for all Americans safe, healthful, productive, and esthetically and culturally pleasing surroundings;

(3) attain the widest range of beneficial uses of the environment without degradation, risk to health or safety, or other undesirable and unintended consequences;

(4) preserve important historic, cultural, and natural aspects of our national heritage, and maintain, wherever possible, an environment which supports diversity, and variety of individual choice;

(5) achieve a balance between population and resource use which will permit high standards of living and a wide sharing of life's amenities; and

(6) enhance the quality of renewable resources and approach the maximum attainable recycling of depletable resources.

(c) The Congress recognizes that each person should enjoy a healthful environment and that each person has a responsibility to contribute to the preservation and enhancement of the environment.

SEC. 102

The Congress authorizes and directs that, to the fullest extent possible: (1) the policies, regulations, and public laws of the United States shall be interpreted and administered in accordance with the policies set forth in the Act, and (2) all agencies of the Federal Government shall—

(A) utilize a systematic, interdisciplinary approach which will insure the integrated use of the natural and social sciences and the environmental design arts in planning and in decisionmaking which may have an impact on man's environment;

(B) identify and develop methods and procedures, in consultation with the Council on Environmental Quality established by title II of this Act, which will insure that presently unquantified environmental amenities and values may be given appropriate consideration in decisionmaking along with economic and technical considerations:

(C) include in every recommendation or report on proposals for legislation and other major Federal actions significantly affecting the quality of the human environment, a detailed statement by the responsible official on—

(i) the environmental impact of the proposed action,

(ii) any adverse environmental effects which cannot be avoided should the proposal be implemented,

(iii) alternatives to the proposed action,

(iv) the relationship between local short-term uses of man's environment and the maintenance and enhancement of long-term productivity, and

(v) any irreversible and irretrievable commitments of resources which would be involved in the proposed action should it be implemented.

Prior to making any detailed statement, the responsible Federal official shall consult with and obtain the comments of any Federal agency which has jurisdiction by law or special expertise with respect to any environmental impact involved. Copies of such statement and the comments and view of the appropriate Federal, State, and local agencies, which are authorized to develop and enforce environmental standards, shall be made available to the President, the Council on Environmental Quality and to the public as provided by section 552 of title 5, United States Code, and shall accompany the proposal through the existing agency review processes;

(D) study, develop, and describe appropriate alternatives to recommended courses of action in any proposal which involves unresolved conflicts concerning alternative uses of available resources;

(E) recognize the worldwide and long-range character of environmental problems and, where consistent with the foreign policy of the United States, lend appropriate support to initiatives, resolutions, and programs designed to maximize international cooperation in anticipating and preventing a decline in the quality of mankind's world environment;

(F) make available to States, counties, municipalities, institutions, and individuals, advice and information useful in restoring, maintaining, and enhancing the quality of the environment;

(G) initiate and utilize ecological information in the planning and development of resource-oriented projects; and

(H) assist the Council on Environmental Quality established by title II of this act.

SEC. 103

All agencies of the Federal Government shall review their present statutory authority, administrative regulations, and current policies and procedures for the purpose of determining whether there are any deficiencies or inconsistencies therein which prohibit full compliance with the purposes and provisions of this Act and shall propose to the President not later than July 1, 1971, such measures as may be necessary to bring their authority and policies into conformity with the intent, purposes, and procedures set forth in this Act.

SEC. 104

Nothing in section 102 and 103 shall in any way affect the specific statutory obligations of any Federal agency (1) to comply with criteria or standards of environmental quality, (2) to coordinate or consult with any other Federal or State agency, or (3) to act, or refrain from acting contingent upon the recommendations or certification of any other Federal or State agency.

SEC. 105

The policies and goals set forth in this Act are supplementary to those set forth in existing authorizations of Federal agencies.

Appendix E
Units,
Conversion Factors,
Physical Constants, and
Useful Numerical Data

Definitions of Units of Energy

Joule
The mks unit of energy. A mass of 1 kilogram moving with a velocity of 1 meter/sec has a kinetic energy ($1/2\ mv^2$) of 1/2 joule. Similarly, if a force of 1 newton is exerted through a distance of 1 meter, the energy expended (work done) is 1 joule. Abbreviated J.

Erg
The cgs unit of energy. A mass of 1 gram moving with a velocity of 1 cm/sec has a kinetic energy of 1/2 erg. Similarly, if a force of one dyne is exerted through a distance of 1 cm, the energy expended (work done) is 1 erg. 1 erg = 10^{-7} joules.

Foot-poundal
A unit of energy defined analogously to the joule and erg, in the "feet-pounds-seconds" system of units. A mass of 1 pound moving with a velocity of 1 ft/sec has a kinetic energy of 1/2 foot-poundal. Similarly, if a force of 1 poundal is exerted through a distance of 1 foot, the energy expended (work done) is one foot-poundal. (A poundal is the force required to accelerate a mass of 1 pound by 1 ft/sec^2.) Abbreviated pdl-ft.

Foot-pound
If a force equal to that exerted by gravity on a mass of 1 pound (at the earth's

surface) is exerted through a distance of 1 foot, the energy expended is 1 foot-pound. For example, the energy required to raise a mass of 1 pound by 1 foot (at the earth's surface) is 1 foot-pound. Abbreviated ft-lb.

Kilogram-meter

A unit of energy defined analogously to the foot-pound in the mks system. The energy required to raise a mass of 1 kilogram by 1 meter is 1 kilogram-meter. Abbreviated kgf-m. Also known as meter-kilogram (m-kgf).

Calorie

Originally a unit of heat energy. Calories come in two sizes: kilocalories (kcal) and gram-calories (g-cal or cal). One kilocalorie is the amount of energy required to raise the temperature of 1 kilogram of water by 1°C (from 15°C to 16°C). Also known as kilogram-calorie (kg-cal). One gram-calorie is the amount of energy required to raise the temperature of 1 gram of water by 1°C.

British thermal unit

The unit analogous to the calorie in the British system of units. One Btu is the amount of energy required to raise the temperature of one pound of water by 1°F.

Kilowatt-hour

A unit of energy defined in terms of power × time. If energy is expended at a rate of 1 kilowatt for a duration of 1 hour, then a total of 1 kilowatt-hour of energy has been expended. Abbreviated kWh or kW-hr.

Horsepower-hour

A unit defined analogously to the kilowatt-hour, but with the horsepower as the unit of power. If energy is expended at a rate of 1 horsepower for a duration of 1 hour, then a total of 1 horsepower-hour of energy has been expended. Abbreviated hph or hp-hr.

Electron-volt

A unit of energy useful in considering atomic, molecular, and nuclear processes. The work done or energy transmitted when an electric charge equal to the charge of one electron falls through the electric potential of 1 volt. Abbreviated eV.

Fuel units

By these units (barrels of oil, tons of coal, and 1000's of cubic feet of natural gas), one means the available energy content of the unit of fuel.

Conversion Among Energy Units

Table E.1 gives the conversion factors among the most commonly used units of energy.

Table E.1 Relations among several units of energy.

	is equal to					is the energy content of		
	joules	kcal	Btu	kWh	eV	natural gas	bbl oil	coal
1 joule	1 joule	2.39×10^{-4} kcal	9.48×10^{-4} Btu	2.78×10^{-7} kWh	6.25×10^{18} eV	0.88×10^{-6} cu ft natural gas	1.6×10^{-10} bbl oil	0.35×10^{-10} tons coal
1 kilocalorie	4186 joules	1 kcal	3.97 Btu	1.16×10^{-3} kWh	2.62×10^{22} eV	3.7×10^{-3} cu ft natural gas	6.8×10^{-6} bbl oil	1.5×10^{-6} tons coal
1 British thermal unit	1055 joules	0.252 kcal	1 Btu	2.93×10^{-4} kWh	6.59×10^{21} eV	0.93×10^{-3} cu ft natural gas	1.72×10^{-7} bbl oil	0.37×10^{-7} tons coal
1 kilowatt-hour	3.6×10^{6} joules	860 kcal	3413 Btu	1 kWh	2.25×10^{25} eV	3.2 cu ft natural gas	5.9×10^{-4} bbl oil	1.3×10^{-4} tons coal
1 electron-volt	1.60×10^{-19} joules	3.82×10^{-23} kcal	1.52×10^{-22} Btu	4.45×10^{-26} kWh	1 eV	1.4×10^{-25} cu ft natural gas	2.6×10^{-29} bbl oil	5.6×10^{-30} tons coal

	has an energy content of					has the same energy content as		
	joules	kcal	Btu	kWh	eV	natural gas	bbl oil	coal
1000 cu ft natural gas	1.13×10^{9} joules	2.7×10^{5} kcal	1.07×10^{6} Btu	314 kWh	7.05×10^{27} eV	1000 cu ft natural gas	0.184 bbl oil	0.041 ton coal
1 barrel oil (42 U.S. gal)	6.12×10^{9} joules	1.46×10^{6} kcal	5.8×10^{6} Btu	1700 kWh	3.82×10^{28} eV	5400 cu ft natural gas	1 bbl oil	0.22 ton coal
1 metric ton average coal	2.7×10^{10} joules	6.55×10^{6} kcal	2.6×10^{7} Btu	7620 kWh	1.71×10^{29} eV	24,000 cu ft natural gas	4.5 bbl oil	1 ton coal

Definitions of Units of Power

Any unit of energy per unit time is a legitimate unit of power, and almost all conceivable combinations have been used. Thus we have cal/sec, Btu/hr, foot-pound/min, etc. It these cases, the name of the unit is sufficient to define it. Two units whose definitions are not evident from their names follow.

Watt
The mks unit of power defined as one joule per second. Also in frequent use are the kilowatt (kW), which is 1000 watts, and the megawatt (MW) which is one million watts.

Horsepower
Originally intended as the power capabilities of a horse. One horsepower is defined as 550 foot-pounds/sec, and equals 0.746 kilowatts.

Physical Constants and Useful Numerical Data

The sorts of numerical data relevant to the considerations of this book are so varied and diverse, and so extensive, that any tabulation will be incomplete. I give here a very brief collection. It may reduce, but will certainly not eliminate, your need to consult more extensive tabulations.

Item	Symbol	Value
Velocity of light	c	3.00×10^8 m/sec
Planck's constant	h	6.63×10^{-34} joule-sec
Stefan-Boltzmann constant	σ	0.567×10^{-11} joule cm^{-2} deg^{-4} sec^{-1}
Boltzmann's constant	k	1.38×10^{-23} joules/deg
Acceleration of gravity at earth's surface	g	9.81 m/sec^2
Avogadro's number	N	6.02×10^{23} atoms/gm-mole
Solar constant	S	1.36 kW/m^2
Heat of vaporization of water		~ 585 kcal/kg
Area of earth (70% ocean, 30% land)		2×10^8 sq mi, 1.3×10^{11} acres
Area of the United States (including Alaska)		3.6×10^6 sq mi, 2.3×10^9 acres
Age of:		
Earth		$\sim 4.5 \times 10^9$ yrs
Life on earth		3×10^9 yrs
Mammals on earth		150×10^6 yrs
Homo sapiens		1 to 2×10^6 yrs
Agriculture		9000 yrs
Civilization		6000 yrs
Industrial period		200 yrs
Population figures, 1972		
World		3.7×10^9 people; growth rate 2.0%/yr
United States		209×10^6 people; growth rate 1.1%/yr
Human power use, 1970		
World		6×10^{12} W; 1.5×10^3 watts per capita
United States		2×10^{12} W; 10^4 watts per capita

Appendix F
Answers to Selected Problems

1.4

Flux at the screen is 0.093 watts/m²; energy/day is 2.4×10^4 joules. (*Note:* These answers assume that the energy leaving the light bulb by conduction and convection is negligible.)

1.5

The 64th square required 1.8×10^{11} tons.

1.7

The supply lasts 301 years; 50 years from the end, an apparent 135-year supply remains.

2.2 (b)

$C_A = 3740$ cal cm²-°C (240 air; 3500 oceans, 1.2 soil); $\tau = 295$ days (19 air, 276 oceans, 0.1 soil).

2.5

262°K.

2.7

$T(z) = [(S/8\sigma)(1 + (L - z)/\lambda)]^{1/4}$; however, the temperature of the earth's surface is *not* equal to $T(0)$, because conduction has been neglected. T (surface) $= [(S/8\sigma)(2 \times L/\lambda)]^{1/4}$.

3.6

8.5 percent.

3.8

Fewer than 40 million people; less than 1 percent of caloric needs (using < 1 g/m²-day from Table 3.1).

4.2
0.75 ton coal, 3.4 barrels oil, 18,400 ft³ natural gas; 40 m² if solar energy is used at the same efficiency as fossil fuels (400 m² if efficiency is 1/10 that of fossil fuels).

4.5
0.12 m/hr.

4.8 (a)
World oil and gas, 48 years; world coal, 110 years; U.S. uranium with breeding, 200 years; world deuterium, 620 years.

4.8 (b)
297 years.

5.2 (a)
23.6 fissions in 10th generation.

5.2 (b)
145 generations for 10^6 increase.

5.6 (a)
$$\frac{d^2P}{dt^2} - \frac{1}{P}\left(\frac{dP}{dt}\right) + \frac{P\beta\alpha}{\tau}(P - P_0) = 0.$$

5.10 (a)
92 percent for one layer of glass; $\sim$100 percent for selective surface with $\epsilon_{IR} = 0.05$.

5.10 (b)
18 percent for one layer of glass; 39 percent for selective surface with $\epsilon_{IR} = 0.05$. (These answers assume a solar flux of 1 kW/m².)

5.12 (b)
52 percent.

6.3 (b)
For a flow of 10,000 ft³/sec, $x_0 = 134$ miles; $\Delta T(0) = 1.7°C$.

6.4
Hypolimnion temperature rise of 0.58°C; hypolimnion conversion 1.17 m.

6.5 (c)
Penetration $= 1.14$ m.

6.11
Dose is 0.020 rad/hr; 0.026 R/hr.

6.12
Approximately 240 ion pairs per cell nucleus.

6.15
Equilibrium levels are 17,000 Ci of tritium, 2000 Ci of ^{133}Xe.

7.1 (c)
Doubling times shrank to 70 years for world population around 1925, for world per-capita energy use around 1930, and for total world energy use around 1860.

7.2 (b)
The first epoch lasted 9×10^5 years; the eleventh will take 10 years.

7.12

For the "typical" well-heated house (Table 8.2), 80 percent gas furnace efficiency, 5050 deg-day/yr, direct costs are $121 (gas) and $277 (electricity). (Installation and maintenance costs are not included.)

8.7

3.9 percent fuel saving, assuming $\alpha = 2.5\%$/hr.

8.14

Fuel requirements grow 5.2%/yr; waste heat grows 4.0%/yr.

Index

Index